Genetic Tools for Improving Livestock Performance

NIPA® GENX ELECTRONIC RESOURCES & SOLUTIONS P. LTD.
New Delhi-110 034

About the Editors

Dr. Deepak Sharma, a Professor & Head at the Department of Animal Genetics and Breeding, Co.V.Sc & A.H, DUVASU, Mathura, U.P., is a seasoned professional has more than 18 years of teaching and research experience in the field of Molecular Genetics and Population Genetics. With academic credentials from CSAUAST, Kanpur and IVRI, Izatnagar Dr. Sharma's expertise is evident in his extensive publication record, including 48 research articles, 10 book chapters, and 6 practical manuals. He has supervised numerous Masters and Ph.D. students, with a total of 6 Masters and 2 Ph.D. students under his guidance. Additionally, Dr. Sharma has made significant contributions to both externally and university-funded projects. His dedication to scientific animal husbandry, coupled with leadership in academic roles, underscores his commitment to excellence and innovation in the field. Dr. Deepak Sharma's contributions make him an invaluable asset to DUVASU and the broader scientific community.

Dr. Satyendra Pal Singh, an Associate Professor at the Department of Animal Genetics and Breeding, Co.V.Sc & AH, DUVASU, Mathura, U.P., has been instrumental in advancing Molecular Genetics and Population Genetics research at the institution and has more than 14 years of teaching and research experience. With academic qualifications from renowned institutions like GBPUAT & AH, Pantnagar, GADVASU, Ludhiana, and IVRI, Izatnagar. Dr. Singh's expertise lies in Molecular Genetics & Population Genetics. He has contributed significantly to the field, with a notable publication record comprising 42 research articles, 8 book chapters, and 3 review articles. Additionally, Dr. Singh has actively participated in projects, both externally and university-funded, furthering scientific knowledge in his area of specialization. Overall, his commitment to research, education, and professional development underscores his contribution to the field of Animal Genetics and Breeding at DUVASU, Mathura.

Rakesh Goel, M.Phil (Statistics), Assistant Professor Department of Animal Genetics and Breeding, College of Veterinary Science and Animal Husbandry, U.P. Pandit Deen Dayal Upadhyaya Pashu Chikitsa Vigyan Vishwavidyalaya Evam Go-Anusandhan Sansthan, Mathura, has more than 30 years of teaching and research experience in the field of Computer & Biostatistics. He is associated with the research activities of the students, teachers and research scholars of Veterinary and Animal Sciences, Medical Sciences, Agricultural Sciences and Animal Biotechnology. He has also organized several computer trainings. He is also the author of "Question Bank of Statistics" written to meet the requirements of various competitive examinations and another book of "Statistics and Computers for Animal and Veterinary Sciences" written to meet the requirements of the undergraduate students of Veterinary and Animal Science, Animal Biotechnology and other related fields.

Dr. Avneesh Kumar, an Assistant Professor at DUVASU, Mathura, holds a B.V.Sc & A.H. from DUVASU and an M.V.Sc. from IVRI, Izatnagar. He served as a Veterinary Officer in the State A.H. Dept., Uttar Pradesh, before joining DUVASU in May 2017. Dr. Kumar is a member of various professional organizations, including the Society for Biosafety and SOCDAB. His contributions to research include book chapters, Genbank submission, and multiple research papers. With active participation in seminars and workshops, he enriches academic discourse in Animal Genetics & Breeding.

Genetic Tools for Improving Livestock Performance

Deepak Sharma
Professor & Head
Animal Genetics and Breeding
College Of Veterinary Science and Animal Husbandry
Uttar Pradesh Pandit Deen Dayal Upadhyaya Pashu Chikitsa
Vigyan Vishwavidyalaya Evam Go Anusandhan Sansthan
Mathura-281 001, Uttar Pradesh, India

Satyendra Pal Singh
Associate Professor
Animal Genetics and Breeding
College Of Veterinary Science and Animal Husbandry
Uttar Pradesh Pandit Deen Dayal Upadhyaya Pashu Chikitsa
Vigyan Vishwavidyalaya Evam Go Anusandhan Sansthan
Mathura-281 001, Uttar Pradesh, India

Rakesh Goel
Assistant Professor
Animal Genetics and Breeding
College Of Veterinary Science And Animal Husbandry
Uttar Pradesh Pandit Deen Dayal Upadhyaya Pashu Chikitsa
Vigyan Vishwavidyalaya Evam Go Anusandhan Sansthan
Mathura-281 001, Uttar Pradesh, India

Avneesh Kumar
Assistant Professor
Animal Genetics and Breeding
College Of Veterinary Science And Animal Husbandry
Uttar Pradesh Pandit Deen Dayal Upadhyaya Pashu Chikitsa
Vigyan Vishwavidyalaya Evam Go Anusandhan Sansthan
Mathura-281 001, Uttar Pradesh, India

NIPA® GENX ELECTRONIC RESOURCES & SOLUTIONS P. LTD.
New Delhi-110 034

NIPA® GENX ELECTRONIC
RESOURCES & SOLUTIONS P. LTD.

101,103, Vikas Surya Plaza, CU Block
L.S.C.Market, Pitam Pura, New Delhi-110 034
Ph : +91 11 4386 0225, 9717133558, 9540816132
E-mail: newindiapublishingagency@gmail.com
Website: www.niparesources.com

Print ISBN: 978-81-19103-04-1

ebook ISBN: 978-81-19103-05-8

Composed and Designed by NIPA®.

उ.प्र.पं. दीनदयाल उपाध्याय पशु चिकित्सा विज्ञान विश्वविद्यालय
एवं गो अनुसंधान संस्थान, मथुरा-281001 (उ.प्र)

U.P. Pandit Deen Dayal Upadhyaya Pashu Chikitsa Vigyan Vishwavidyalaya Evam Go Anusandhan Sansthan (DUVASU) Mathura-281001 (U.P.)

प्रो.(डॉ.) ऐ.के. श्रीवास्तव
कुलपति
Prof. (Dr.) A.K. Srivastava
Vice-Chancellor

Foreword

The present book entitled "Genetic Tools for Improving Livestock Performance", has been well prepared to meet the requirements of a comprehensive overview of the genetic tools and techniques that can be employed to improve the performance of livestock. It covers a wide range of topics, including genomics, genetic selection and breeding strategies for the livestock

I am sure that the book serves as a reference book not only for students in the field of Animal Genetics and Breeding but also for those who want to know or who need to be familiar with advances in genetics, genomics and biotechnology,

I am really very happy to see that this book has been designed in accordance with the core concept of the topic and will be an invaluable resource for the students, teachers and researchers.

I appreciate the efforts made by the authors who are experts in their respective field have provided valuable insights and perspectives on the latest developments in the rapidly evolving field. I congratulate the Department of Animal Genetics and Breeding for bringing out the book in the present form and wish them all the success in their future endeavors.

(A.K. Srivastava)

Tel. : 0565-2470199; Fax: 0565-2470819; Mob. 9466592661; Email: duvasuvc@gmail.com
https://www.upvetuniv.edu.in

Preface

It is with great excitement that we present this book, Genetic Tools for Improving Livestock Performance, which presents a comprehensive overview of the application of genomics within the Livestock industry and Animal Husbandry practices in the national scenario Genomics is an incredibly exciting field which is developing at a rapid rate, and this book serves to present some of the latest advancements in the field while also delving into its basic concepts. This book aims to explore the full gamut of genomic advancement in the animal-breeding sector.

The potential use of genomic technology in livestock and poultry breeding is tremendous, greatly enhancing the ability of farmers to produce healthier and more resistant livestock and poultry. By using genetic information, breeders can select animals with specific traits, allowing them to produce higher quality products in a more efficient manner. Consequently, this technology has the potential to revolutionize the industry, resulting in improved yields and profits while also improving animal welfare.

Within these pages, readers will find information on the many new methods that have been developed to analyze genetic markers and to identify desired traits in animals. The book is aimed at a wide range of readers, from those with a basic understanding of genetics to those with more detailed knowledge as both the basics of genomics and its more advanced applications, such as the application of genomics to the identification of specific traits in breeds and the development of new breeds etc. has been covered in this book. By examining the areas of genotyping, SNP (single nucleotide polymorphism) analysis, and high-density direct genomic selection (DNA-based selection), readers will gain a better understanding of how genomic technology can be utilized within the field.

Authors have also tried to provide further insights into how genomic advancements have been applied within the national scenario. We discuss the numerous methods used for animal-breeding research, including classical livestock selection, association studies, and genomic-enabled selection. Furthermore, the implications of this technology for different stakeholders particularly farmers and breeders, are addressed.

Ultimately, this book is intended to provide readers with a greater understanding of the potential of genomics in the field of animal breeding. This book includes contributions from specialists in the field and will prove very helpful to researchers, industry professionals, and students seeking to gain insight into the latest genetic tools for improving livestock and poultry production.

Editors

Contents

Part B: Improving the Performance of Livestock Using Genetic Tools

Part A: Genetic Tools

1

Genomics New Insights on Animal Reproduction and Production

Satyendra Pal Singh, Avneesh Kumar and Deepak Sharma

Department of Animal Genetics and Breeding, CoVSc & AH, DUVASU
Mathura, Uttar Pradesh

Genomics is the study of whole genomes of organisms, it uses a combination of DNA sequencing methods, bioinformatics to sequence, assemble, analyse the structure and function of genomes. In another words, it can be defined as it is the study of the structure, function, and inheritance of the genome of an organism. Moreover, genomics focuses on interactions between loci and alleles within the genome and other interactions such as epistasis, pleiotropy and heterosis. Genomics is divided into two basic areas: structural genomics, characterizing the physical nature of whole genomes; and functional genomics, characterizing the transcriptome (the entire range of transcripts produced by a given organism) and the proteome (the entire array of encoded proteins).

Most biological traits are controlled by a large number of genes, which, in addition to their additive effects, may interact with each other, and their expression might be altered based on a variety of environmental effects. Researchers who were among the first to use molecular genetics to the study of cattle began by employing DNA markers to locate genes or sections of the genome affecting features of interest. For instance, identifying a group of genes based on Genome wide association study (GWAS) provides evidence that a causal relationship between a phenotype and a single nucleotide polymorphisms (SNPs) may exist. These findings of GWAS, however, do not improve our understanding of complex traits in which many pathways may be involved. The growing use of genotyping technologies has generated a large quantity of genomic information on the dairy population. Since the draft sequence of the bovine genome was published in 2009 (Zimin et al., 2009), whole-genome sequence data of cattle have also become available. The emergence of genomics was initially slow, but has exploded in the last decade. Fuelled by decreasing costs, advances in next-generation sequencing (NGS) technologies enable identification of more complex forms of genetic variation (e.g. short insertions and deletions (INDELs), copy number variations (CNVs), precise genome-wide linkage disequilibrium patterns (Wang et al., 2013) than previously possible.

Application of Genomics

Sequencing the genomes of animals could be beneficial to animal production practices, animal health and welfare, and to our understanding of the genetic basis of diseases in both animals and humans. Beyond these more applied areas of study, sequencing genomes also presents opportunities for increasing our basic knowledge of the evolutionary pathways of these and related species.

The traits of interest are primarily production traits. These include growth (both prenatal and postnatal), reproduction, egg production, and carcass traits, including fat deposition within and between the muscles, inside organs and layered under the skin. Other production traits are meat tenderness and palatability, as well as milk production. Once a particular animal genome is sequenced, it might be possible to determine which gene or genes affect a trait and thus give breeders the information they need to enhance the production traits of that animal.

Animal Breeding

For thousands of years, these livestock have been bred for desirable traits including disease resistance and rapid growth in farm animal. Animal breeders have been very successful in their manipulation of the genomes of livestock species for centuries, making use of the fact that natural variations exist within a species, within a breed, and within a population. This has allowed animal breeders to create livestock with desirable characteristics. Conventional breeding has been conducted without molecular understanding of the genes governing a quantitative trait locus. The precision of traditional, selective breeding is low and genetic change is poorly characterized. Theoretically, with information from a sequenced genome, it will be possible to have much more precision with breeding efforts and even to genetically engineer specific traits by adding, removing, or altering individual genes.

Pedigree information and dairy production recording programs provided the initial datasets for estimation of genetic merit and improvement of dairy breeds. Major developments also occurred with the introduction of selection index theory (Hazel, 1943) and best linear unbiased prediction (BLUP) methodologies. Substantial genetic gains were made using these breeding strategies in many areas, including high heritability milk production traits. Although the combination of extensive pedigree data and phenotypes enhanced selection programs, breeding for low-heritability traits, such as fertility and health, and other hard-to-phenotype traits, compelled the exploration of novel methods to achieve increased accuracies.

Breeders have improved production characteristics in their herds by selecting superior individuals as progenitors for future generations. This improvement in "breeding values" was accomplished by integrating phenotypic recording of individual performance with genealogical data. These findings highlight the impressive potential of conventional breeding techniques, but they also highlight the limitations of those techniques under certain conditions, such as when traits are difficult to measure, have low heritability, or cannot be measured quickly, cheaply, and accurately across a large population. Such difficult-to-measure traits are often critically important; they include fertility, longevity, feed efficiency, and disease resistance. Selection for these traits must be achieved through genomic means.

Marker-Assisted Selection to Genomic Selection

In order to eliminate potentially harmful alleles of genes (such halothane in pigs or fish taint in brown-shelled chicken eggs), breeders began using gene marker technology in the form of marker-assisted selection (MAS) in the 1990s (such as the Polled condition in cattle). However, MAS relies on quantitative estimates of relationships between gene alleles and qualities of interest in a given population, which can be difficult to obtain. Accordingly, it needs to start at home. In addition, MAS can account for just a small fraction of the heritable variation observed in humans. However, genomic selection technologies are now paving the way to a paradigm shift in animal breeding, with the availability of cheap whole-genome SNP panels for the key livestock species. With a large enough panel of genetic markers, we can go from MAS to genomic selection, tracking the inheritance of the entire genome rather than just a few regions of interest. Similar performance in animals can now be described without reference to parental ties thanks to the availability of dense sets of single-nucleotide polymorphisms (SNP).

Genomic Selection

The advent of DNA sequencing and high-throughput genomic technologies has resulted in the discovery of a large number of SNP in cattle and other food animal species. The use of dense SNP arrays that encompass the bovine genome so it describe the vast bulk of genetic variation in significant traits has indeed been proposed by an approach known as genomic selection or whole-genome selection. Automated methods for SNP genotyping are now extensively available. Whole-genome selection and genomic selection are both terms that refer to the same concept (Meuwissen et al., 2001). In practice, genomic selection refers to selection decisions based on genomic estimated breeding values (GEBV). These GEBV are calculated by estimating SNP effects from prediction equations, which are derived from a subset of animals in the population (*i.e.,* a reference population) that have SNP genotypes and phenotypes for traits of interest.

Principle of Genomic Selection

Meuwissen et al. (2001) were the first to describe genomic selection. It is rooted in the fundamental concept that details from a large number of markers can be used to estimate breeding values even without having accurate information of where particular genes are located on the genome. Genomic selection had first been discussed by Meuwissen et al. (2001). It is anticipated that there will invariably be an SNP in close proximity to a particular gene or DNA fragment of interest when using tens of thousands of SNP that have been carefully chosen to be representative of the entire genome. The existing linkage disequilibrium in between (or several) SNP and a causal mutation will be substantial, and this can then be used to explain a significant fraction of the variation in the observed trait.

Methodology

Genomic selection begins with a large sample size of animals that have reliable phenotypes for the trait in question, known as a reference or training population (s). The use of a whole-genome SNP array for genotyping in this group is recommended as well. The collected information will be used as a benchmark against which a statistical model may be constructed, with the goal of assessing the influence of each SNP on the trait(s) in question. As a result, we can use this prediction equation to determine the genetic estimated breeding value (GEBV). Following validation, the prediction equation can be used to estimate the genomic breeding value of uncharacterized animals given only their SNP array genotypes, in the lack of reliable information on their phenotypes. The heritability of the variable being evaluated and the population size both affect the GEBV's precision. In the chicken, for example, González-Recio et al. (2009) referenced a 4-fold increase in GEBV accuracy over the parent average for feed conversion efficiency.

Genomic selection allows AI companies to make decisions based on GEBV, which are available at a very young age. Therefore, younger bulls can be used as sires of sons in the SM pathway (Sire of male pathway), and the age at which they can be used is limited only by their sexual maturity. Instead of waiting a minimum of 4.5 years to use progeny-tested bulls as sires of sons, AI companies can use the best DNA-tested young bulls as sires of sons by roughly 1 year of age. This drastically reduces the generation interval in the SM pathway and, as noted by Schaeffer (2006), it could lead to doubling of the rate of genetic gain. Yearling bulls that have GEBV information but lack phenotypic data on their daughters are often referred to as "genomic bulls." There has been an immense shift among the AI companies toward the use of genomic bulls in the past 3 years. Some AI companies use almost all genomic bulls as sires of sons, whereas other companies use a combination of genomic bulls and progeny-tested bulls. Genomic bulls that are considered as sires of sons by AI companies have higher genetic merit, on average, but lower average reliability, so AI companies try to minimize risk by considering a larger number of genomic bulls as sires of sons.

Selection Intensity

Selection intensity is dependent on the size of the population and on the proportion of animals selected from the population to serve as parents of the next generation. Greater selection intensity can be achieved in large populations because more selection candidates can be screened in search of genetic "outliers." In the era of genomic selection, DNA samples can be taken shortly after birth, and AI companies can screen thousands of bull calves to find a few elite individuals for their breeding programs. Potential bull dams can be tested as well, and this gives AI companies and breeders an opportunity to screen thousands of potentially elite cows and heifers on commercial farms for the purpose of identifying a few superior individuals that can be propagated by MOET or IVF.

Genetic Variation

The genetic standard deviation (σA) reflects the underlying genetic variability of a given trait within the population. Inbreeding decreases the effective population size, which can reduce the amount of genetic variation available for selection and reduce the rate of genetic progress. However, in comparison with other factors in the equation for genetic progress, relatively little can be done to increase the amount of genetic variation within a population.

Advantages

- Implemented very early in life.
- Not sex limited.
- Can be extended to any traits that are recorded in a reference population.
- Provides for difficult-to-improve traits.
- Better selection accuracy.
- Reducing the generation interval.
- Increasing the intensity of selection.
- Compares favourably to MAS in terms of the proportion of genetic variation that it can explain.
- Not restricted to any particular family.

Present Status of Genomic Selection in Livestock

The accuracy of genomic prediction exceeds 0.80 for production traits and 0.70 for fertility, longevity, somatic cell count, and other features in dairy cattle (Wiggans et al., 2011). Because most of the animals in the reference populations are progeny-tested bulls with highly accurate phenotypes based on average daughter performance, these high accuracies are a reflection of the large reference populations for each breed that have been assembled to enable genomic predictions. In addition, these high accuracies reflect the fact that many of the animals in the reference populations have been assembled to enable genomic predictions. In addition, the GEBV are frequently utilised to make predictions on the close relatives of animals that are included in the reference population. Collaboration between countries to assemble these large reference sets is one of the characteristics of dairy genomic predictions. To facilitate this collaboration, three consortiums have been established: Eurogenomics, which includes the Netherlands, Germany, France, the Nordic countries, Spain, and Poland; The North American Consortium, which includes the United States of America, Canada, Italy, and Great Britain; and a "rest of the world" consortium, which consists of a number of remaining countries. For the goal of genomic prediction, the genotypes of about 2 million dairy cattle have now been collected from throughout the globe. Only in the United States of America have genotyping efforts been focused on a total of 934,780 Holstein animals, 120,439 Jersey animals, 19,588 Brown Swiss animals and 4,767 Aryshire animals (Wiggans, personal communication; https://www.cdcb.us/Genotype/cur density.html). Genotyping has been performed on comparable numbers of animals in a number of other nations combined, including 360,000 in France alone (Boichard, personal communication). The application of genomic selection in dairy cattle has led to an increase in genetic gain, which has now been confirmed by genetic trend analysis in a number of nations. This genetic gain has been passed down to future generations. For instance, with the implementation of genomic selection, the rate of genetic gain has almost doubled in Canada (VanDoormal, personal communication). There is also some evidence to show that genetic selection has contributed to an increase in the annual rate of inbreeding (Schenkel, 2012). Therefore, one of the most important topics for research in the future will be how to maximise the genetic benefit via genomic selection while simultaneously limiting the rate of inbreeding.

Challenges

- Large training populations are required to obtain acceptable accuracies for breeding values
- The rate of inbreeding per year will increase.
- The need of continuous phenotyping: Some simulation experiments have shown that accuracy using the same markers is rapidly lost generation after generation and new training populations are required.
- Preselecting bulls in dairy cattle using genomic information can lead to biased predictions with lower accuracy, leading to a decrease in genetic progress and distorting international dairy bulls comparisons
- SNP estimates calculated from a Holstein-Friesian reference population did not produce accurate GEBV in Jersey bulls, and the correlations ranged from −0.1 to 0.3 when the SNP effects from one breed were used to calculate GEBV in another breed vice versa.
- Implanting GS in current breeding schemes

Future Directions & Conclusion

The ability to investigate the genome, transcriptome, epigenome, and metagenome of any species is opening a new world of possibilities. All economically impactful agricultural species, subspecies, and their pathogens, thousands of related genomes will also be sequenced in the near future which gives us critically important information for the implementation of genomic selection programs. Genomic selection combined with (MOET) could greatly decrease generation length, may be used to pick the best embryos for next generation. We could train the prediction equation on commercial animals (often crossbreds, run under a harsher environment), rather than stud animals (purebred stud Animals).

References

González-Recio, O., D. Gianola, G. J. Rosa, K. A. Weigel, and A. Kranis. 2009. Genome-assisted prediction of a quantitative trait measured in parents and progeny: Application to food conversion rate in chickens. Genet. Sel. Evol. 41:3.

Hazel, L., N. 1943. The genetic basis for constructing selection indexes. Genetics. 28(6): 476-490.

Meuwissen, T. H. E., B. J. Hayes, and M. E. Goddard. 2001. Prediction of total genetic value using genome-wide dense marker maps. Genetics 157:1819–1829.

Schaeffer, L. R. 2006. Strategy for applying genome-wide selection in dairy cattle. J. Anim. Breed. Genet. 123:218–223.

Schenkel, F. 2012. Inbreeding using genomics and how it can help. World Holstein Federation Conference,Toronto.http://www.whff.info/documentation/documents/2012_03_FlavioSc henkel.pdf.

Wang, C., D. Habier, B.L. Peiris, A. Wolc, A. Kranis, K.A. Watson, S. Avendano,D.J. Garrick, R.L. Fernando, S.J. Lamont, and J.C.M. Dekkers. 2013. Accuracy of genomic prediction using an evenly spaced, low-density single nucleotide polymorphism panel in broiler chickens. Poult. Sci. 92:1712–1723.

Wiggans, G.R., P.M. Vanraden, and T.A. Cooper. 2011. The genomic evaluation system in the United States: Past, present, future. J. Dairy Sci. 94(6):3202–3211.

Zimin, A., V., Delcher, A., L., Florea, L., Kelley, D., R. et al. 2009. A whole assembly of the domestic cow, Bos Taurus. Genome Biol. 10:R42.

2

Genomics A New Breeding Paradigm Shift in Farm Animal Genetic Improvement

Sulochana Sen[1], Rajesh Kumar Vandre[1], Amit Kumar Jha[1] Abhilasha Singh[2] Suman Sant[3] and Anita Tewari[4]

[1]Dept of Animal Genetics and Breeding
[2]Department of Animal Nutrition
[3]Department of Veterinary Extension
[4] Department of Veterinary Public Health and Epidemiology
College of Veterinary Science and A.H., Rewa, NDVSU Jabalpur, Madhya Pradesh

Genetic improvement of farm animal species is accelerating in developed countries with adoption of Genomic selection (GS) due to availability of complete reference genome sequence i.e. gene sequences that can be used to compare and study animal genetic traits. GS has been considered most promising for genetic improvement of the complex traits controlled by many genes (Quantitative traits) each with minor effects as using the conventional methods of selection and breeding the rate of genetic gain per generation is very slow with respect to human demand for livestock products. With this method, animals or plants can be selected for selective breeding on the basis of their genetic merit predicted by whole genome molecular markers spanning over the entire genome. Genomic selection may allow the identification of superior individuals for traits such as increased disease resistance which are not currently considered in animal breeding plans because of technical difficulties. The method can predict the genetic merit of young bulls with greater accuracy as compared to convention methods of genetic merit estimation. Genetic gain (ΔG) in animal breeding programs depends upon the intensity of selection (i), accuracy of predictions (r), genetic variance ($\sigma^2 g$) and generation interval (IG): $\Delta G = i * r * \sigma 2g / IG$ (Ibañez Escriche & Gonzalez-Recio, 2011). Genomic selection increases the rate of genetic gain by decreasing generation interval (Meuwissen & Goddard, 2010) and increasing accuracy of prediction (Hayes *et al.*, 2009). of breeding and genotypic values, reducing cost and accelerating genetic gain by shortening generation interval than conventional progeny testing of bulls (up to double the rate of genetic gain), or identifying animals with greater disease resistance to endemic diseases.

Genomic selection has many benefits, the most important of which are that it may be used very early in an individual's life, that it does not discriminate based on sex, and that it can be extended to any qualities that are recorded in a reference population. If a trait has a low heritability, such as fitness traits (reproductive and survival), or if a trait is difficult to measure, such as fertility, longevity, or feed efficiency; or if a trait cannot be inexpensively, quickly, or correctly measured in a large number of animals; then that trait can be more accurately targeted for selection and improvement using genomic selection and sustainable management of the population (Eggen André, 2012). Moreover, the genomic selection offers ample opportunities for assessment of comprehensive genome wide diversity in population and across breeds and analyzes genetic variation a step ahead, identified within and between breeds using selection and mating practices. The conventional breeding data helps us to identify inbreeding due to identical by descent whereas the genomics can identify identical by state as well which is more précised way to identify inbreeding percentage in a population. Genomic approaches help in locating important populations that need to be preserved, as well as a few locally well-adapted breeds that could be used to breed valuable animals through a combination of selection and cross-breeding. This is made possible thanks to the fact that genomic approaches help identify critical populations.

In European countries due to the availability of reference whole genome sequence of *Bos taurus,* genomic selection has fastened the process of genetic improvement in most of the animal species. Whole genome sequencing and genetic selection are both being used in breeding programmes for important food animals in developed nations (cattle, sheep, goats, chicken, pigs). India being developing country, major challenge in implementing genomic selection is the non-availability of whole genome sequence for indigenous livestock species and cost of genotyping. Several agencies in India are working in this direction and we are in initial phase of developing SNP chip based on Indian species of farm animals which may be available in near future. The use of genomic technologies in developing countries allows for the evaluation of genetic diversity and admixture, as well as signals of selection, in order to locate genomic areas and variants that contribute to variation (Mrode *et al.*, 2019).

First described by Meuwissen *et al.* (2001), genomic selection operates on the premise that single nucleotide polymorphisms (SNPs) are in linkage disequilibrium (LD) with Quantitative trait loci (QTL) in the genome. Thus, by mapping chromosome segments defined by neighbouring SNPs, all QTLs in the genome can be indirectly identified using SNPs as markers (Mrode *et al.*, 2019). The advent in high-throughput single Nucleotide polymorphism (SNP) genotyping technologies such as Microarrays, DNA chips, from Sanger to Next Generation Sequencing (NGS) technologies and Bovine gene map information has made it possible to resequence candidate genes, entire transcriptomes or entire animal genome more efficiently and economically than ever before. In this way, information on thousands of candidate genes and candidate regions can be harnessed for thousands of individuals for sample genetic diversity within and between populations, QTLs mapping, to identify individual genes and to determine their functional diversity.

The development and constant improvement of DNA array technology is a major technological breakthrough which allows for the inexpensive measurement of SNP within a given sample. The success of these DNA arrays is due to the fact that they possess a strong parallel processing capacity, tremendous miniaturization, and a remarkable ability to be automated. DNA array technology was first used for gene expression studies but now are a useful tool for the development of whole genome SNP panels for many species. DNA microarrays are cost effective for screening selected set of SNPs that are easily scalable to large population samples, allowing scientists to perform genome-wide association studies that simply would have been out of reach with traditional (i.e., microsatellite) markers (Eggen André, 2012).

Haplotype information is becoming relevant in functional biology since structural or polymorphic variations can have different phenotypic effects if found on the same or on the homologous chromosome. Haplotype information can be indirectly reconstructed in short regions by population-based phasing approaches that use information on linkage disequilibrium or from next-generation sequencing data using mate-pair information or different insert size libraries that link markers together. The NGS can be applied to resolve the genomic admixture in crossbreeding cattle at the nucleotide level. Genotype imputation which is a process of estimating missing genotypes in terms of the haplotypes and genotypes in a reference panel. It can effectively increase the density of single nucleotide polymorphisms (SNPs) and thus can be widely used in large-scale genome-wide association studies (GWASs) using relatively inexpensive and low-density SNP arrays.

New insights into the growth, nutrition, health, and protection of animals are expected, enabling a better understanding of the molecular mechanisms of traits of interest. Therefore, genomics proposes further opportunities to improve selection accuracy while decreasing the costs, reducing generation intervals, and exploiting new sources of polymorphisms (Dekkers, 2004). Moreover, it would be more effective to integrate GS with other breeding tools and platforms for accelerating the breeding process and thereby further enhancing genetic gain.

Advancement Journey from Marker-Assisted Selection to Genomic Selection in Animal Breeding Improvement

Sax (1923) reported the concept of MAS as an aid to selection using the information from polymorphic loci. Breeders have employed gene marker technology in the form of marker assisted selection (MAS) since the 1990s to eliminate harmful gene alleles (such halothane in pigs or fish taint in brown-shelled chicken eggs) or to select desirable traits based on some marker information (such as the Polled condition in cattle). Ascertaining the relationship between the marker and QTL is essential for MAS. The distance between the genetic marker and the desired qualities determines the strength of the relationship between the marker and the QTL. If a marker is shown to be associated with a QTL, then it can be used in a breeding program's selection procedures. The most common application of marker systems in breeding is MAS, or marker-assisted selection (Raj and Das, 2021). Therefore, the most significant drawback of MAS is that it necessitates prior knowledge about gene alleles or markers that are linked with the qualities of interest, in addition to quantitative estimates of these relationships in the particular population being studied. Therefore, it is necessary for it to be implemented within families. The most significant limitation of MAS is that it can account for just a portion of the genetic differences that exist between people. The reason for this is explained by the fact that for identification of a QTL stringent test for statistical significance are to be performed if these are not applied the tests will be false positives whereas these tests are omitted and effects of all genes are estimated simultaneously (Meuwissen *et al.*, 2001). An extremely high percentage of heritable variation is explained by GS. The GS phenomenon is not exclusive to any one family, unlike the MAS phenomenon. Genomic selection methods and low-cost whole-genome SNP panels for livestock species are ushering in a new era in animal breeding. Changing from MAS to genomic selection is as simple as employing enough genetic markers to track the inheritance of the entire genome as opposed to just a few regions of interest. Despite being ten generations apart, whole-genome sequencing data was able to predict genetic value accurately, with accuracies comparable to those between same-generation test and evaluation data. It is suggested by the authors that high accuracy (genetic/genomic prediction) is possible if genome sequence data, large sample sizes, and a statistical method to detect polymorphisms that are informative are combined (Meuwissen and Goddard, 2010).

Genomic Breeding Value Prediction Model

Principle: Because of the high marker density, each quantitative trait locus (QTL) is in linkage disequilibrium (LD) with at least one or several SNP markers, and a mutation in SNP marker can be strongly linked with a significant fraction of the variation of the observed trait. Genomic selection utilises dense marker (SNP or DNA fragment of interest) maps to anticipate the breeding value of animals. This method is based on the basic principle that each QTL is in LD with at least one or several SNP markers Utilization of a reference population that possesses correct phenotypes for the characteristic at hand is required for the process (s). The SNP array covering the full genome is used to genotype the reference population. Following the completion of the genotyping process, the data obtained will be used as a reference in the development of a statistical model that will estimate the effect that each SNP has on the characteristic or traits of interest. As a result, this will contribute to the development of a prediction equation that can estimate the genomic estimated breeding value (GEBV). Following a step of validation, the genomic breeding value of young selection candidates is determined. Their GEBVs can be computed utilising the prediction equation, the results of which are obtained by adding up all of the relevant SNP effects. This is done in the absence of any precise phenotypes for these animals. Based on their genotypes obtained from the SNP array.

Estimates of the impacts of markers have been obtained through the use of a variety of methodologies, most of which are geared towards simplifying the marker data in some way. Almost all of the models that were reported only contained additive effects. After the impacts of the markers have been evaluated, the breeding values of young candidates for selection can be predicted with an accuracy of up to 0.85, according to reports (Calus, 2010).The main challenge for models that estimate SNP effects is that generally phenotypes are available from a few thousand animals say 1000-2000 animals, while the number of SNP is much larger, for instance, approximately 50000 SNPs for cattle. If one polymorphism is found for each gene we have to calculate gene effect for around 100000 alleles on our breeding goal trait. Our experiment would consist of only 1000-2000 phenotyped and genotyped animals. So estimation 100000 effects from 1000-2000 animals record will not have suffice degree of freedom which in turn will develop practical problem that how this set of equations can be solved. Similarly, problem occurs when we do not know all the genes and we have a high-density marker chip/map to identify all chromosomal segments and we have to estimate effect of all chromosomal segments simultaneously. Thus, reference population that needs to serve multiple lines may optimally contain animals from all lines, and yield nearly as accurate predictions for each of the lines as a line-specific reference population may do. The likely reason for this is that combining animals from different families or lines ensures that only LD that persists across those families or lines is utilized in the prediction equations (Calus, 2010). The problem of huge shortage of degree of freedom i.e. lesser number of observations than the number of independent variables, and the challenge is to find methods that form well in this context . Following prediction models may serve the purpose.

1.Least Squares (LS).

2.Best Linear Unbiased Prediction (BLUP)

3.Bayesian Approach

1. **Least Squares (LS)**: In least-squares analyses, the genes are tested step by step for statistical significance and are added to the model if they significantly improve the fit of the existing model. The effect of the non-significant genes is assumed to be zero. The effect of non-significant genes loci are arbitrary set to zero that are just below the significance threshold and include the full effect of those that are above this threshold. In this method selection of loci with the largest effects results in the selection of overpredicted effects. (Lande and Thompson (1990)) avoided this selection bias by using one-half of the data to select the loci with largest effects and using the other half to re estimate the effects. This splitting of the data set remains a suboptimal use of the information. This method of least square analysis resembles the QTL mapping/Marker Assisted Selection approach. Partial least squares regression (PLS) and sparse PLS are commonly used as alternative to least square analysis method. PLS regression was given by Wold (1966). The number of variables is decreased using partial least squares regression by projecting independent variables onto latent structures. Variable selection and modelling are combined in a single step via sparse PLS (Colombani *et al.*, 2012). In terms of handling big data sets and prediction power, the PLS regression (Wold *et al.*, 2001) method seems to be an effective way to manage genomic selection data. This strategy works best when the predictors' matrix has more variables than observations and when the variables are multicollinear.. (Lê Cao

et al., 2008) in a in a recent method known as sparse PLS regression (sPLS) combined variable selection and modelling in a single phase. By reducing processing time, dimension reduction techniques and variable selection strategies may offer an appealing solution to the increasing number of markers employed in dairy cattle genomic selection (Coster *et al.*, 2010).

2. **Best Linear Unbiased Prediction (BLUP)**: A simple BLUP, as described in Meuwissen *et al.* (2001) and known as genomic BLUP (GBLUP) The simple BLUP assumes that all SNP have an effect sampled from the same normal distribution. GBLUP that fits the allelic effects of each SNP as random effects instead of fixed effect with a normal distribution with known variance. All the allelic effects are estimated simultaneously. When SNP is considered as random effect the variance for all allelic effects has to be estimated in GBLUP so for every gene same variance is used. The GBLUP model assumes that each of the SNPs equally contributes to the additive genetic variance (Meuwissen *et al.*, 2001). The BLUP model can be RRBLUP, ssBLUP. GBLUP has been widely used, with G typically calculated using Van Raden Method 1 (2008). Importantly, the computation of G has made it possible to conduct genetic analyses without pedigree information and estimate the genetic relationship between various groups of animals (Mrode *et al.*, 2018). The fact that genotypic data is only available for a fraction of animals has pushed towards the use of ssGBLUP (Misztal *et al.*, 2009), which combines pedigree and genotypic data to predict genetic merit. This generally leads to higher accuracy because all available data is used, as this usually results in higher accuracy (Cardoso *et al.*, 2014; Silva *et al.*, 2016).

3. **Bayesean approach**: This is identical to GBLUP except that breeding value is estimated using the previous distribution of variance of gene I (Vai), and the variance of allelic effect is supposed to vary for different genes as follows:

   $(V_{ai}) = 0$ with probability p

   $(V_{ai}) \sim \chi^{-2}(v,S)$ with probability (1-p)

   P depends on the mutation rate at the gene, and $\chi^{-2}(v,S)$ denotes inverse-chi squared distribution with v degree of freedom and scale parameter S.

   Marker estimated breeding value (MEBV) is predicted as

   $MEBV_{i=}\sum_j M_{ij}$

where M_{ij} = the estimate of the effect of the j th haplotype of animal i

There are three Bayesean approaches in practice i.e. Bayes A, Bayes B and Bayes C and Bayes **C**$\boldsymbol{\pi}$**.**

Accuracy of Genomic Prediction

The size of the reference population, including the number of animals, markers, and the heritability of the trait under consideration, affects how accurate genomic prediction and GEBV

Nutrition and Animal Genomics

Nutrigenomics and nutrigenetics are new research approaches that strive to optimize health by looking beyond the diet to understand the effects of food at the genetic and epigenetic levels. Nutrigenomics is focused on the effects of diet on health through an understanding of how bioactive chemicals in foods and supplements alter gene expression or the structure of the genome of an animal. Nutrigenetics focuses on how the genetic composition (i.e., genetic variation) of an animal influences their response to a given diet. Results from these studies will aid in formulating nutritionally appropriate diets that may be optimized for animals based on their genomic underpinnings.

Nutrigenomics and nutrigenetics unite many fields: nutrition, bioinformatics, molecular biology, genomics, functional genomics, epidemiology, and epigenomics. The use of multi-disciplinary tools promises new opportunities to investigate the complex interactions of the genome and the diet of an animal. Through these new approaches, the partnerships of the genome and nutrition will be revealed resulting in improved efficiency of diets, enhanced sustainability of animals as a protein source, and improved methods for preventing illnesses.

Conclusion

By incorporating essential biological data and employing effective approaches to connect the knowledge of a statistically related marker locus to a functional gene variant, genomic selection may be enhanced even further. A major barrier for developing countries like India is the absence of evaluation or national testing programmes to determine the quality of germplasms from a genetic standpoint. These programmes record phenotypes on pedigreed animals. We are entering a new era where the data from these genome-wide association studies will be efficiently used in regular testing employing genomic selection, driven by ever-increasing decreases in the cost of quantifying genetic variation. Because genomic selection does not require prior knowledge of alleles or marker placements, it has the potential to be adopted more widely than marker-assisted selection.

A number of innovative biological uses, like targeted exome capture or exome sequencing, Chromatin Immunoprecipitation sequencing (ChIP-Seq), and whole transcriptome shotgun sequencing technology or RNA-Seq have been developed to addresses the cost barrier as well as many biological requirements Exome sequencing prevails over by leaving out intronic sections and only sequencing the exonic regions that would be of more immediate relevance, we can overcome the problem of the high cost of whole genome sequencing. ChIP-Seq is a potent method for examining protein-DNA/RNA interactions and is used to determine the genome-wide binding patterns of a protein of interest, such as a transcription factor. The NGS technologies, RNA-Seq or transcriptome-wide sequencing is utilized to sequence cDNAs from RNA samples. One of the aforementioned methods can reveal differences between several strains or huge populations of related samples.

References

Calus, M. P. L. (2010). Genomic breeding value prediction: methods and procedures. Animal (2010), 4:2, pp 157–164 doi:10.1017/S1751731109991352

Calus, M. P. L. (2010). Genomic breeding value prediction: methods and procedures. Animal (2010), 4:2, pp 157–164 doi:10.1017/S1751731109991352

Colombani C, Croiseau P, Fritz S, Guillaume F, Legarra A, Ducrocq V, Robert-Granié C. A comparison of partial least squares (PLS) and sparse PLS regressions in genomic selection in French dairy cattle. J Dairy Sci. 2012 Apr;95(4):2120-31. doi: 10.3168/jds.2011-4647. PMID: 22459857.

Coster, A., J. W. M. Bastiaansen, M. P. L. Calus, J. A. M. van Arendonk, and H. Bovenhuis. 2010. Sensitivity of methods for estimating breeding values using genetic markers to the number of QTL and distribution of QTL variance. Genet. Sel. Evol. 42:9.

Dekkers, J. C. 2004. Commercial application of marker- and gene-assisted selection in livestock: Strategies and lessons. J. Anim. Sci. 82(E-Suppl.):E313–E328.

Eggen André (2012). The development and application of genomic selection as a new breeding paradigm. Animal Frontiers 2(1): 10-15.

Hayes, B.J., 2009. Invited review: Genomic selection in dairy cattle: progress and challenges. Journal of dairy science, 92(2): pp.433-443.

Ibañez-Escriche, N. & Gonzalez-Recio, O., 2011. Review. Promises, pitfalls and challenges of genomic selection in breeding programs. Spanish Journal of Agricultural Research, 9(2): pp.404-413. Available at: http://revistas.inia.es/index.php/ sjar/article/view/1730.

Lê Cao, K. A., D. Rossouw, C. Robert-Granie, and P. Besse. 2008. A sparse PLS for variable selection when integrating omics data. Stat. Appl. Genet. Mol. Biol. 7:35.

Meuwissen, T., and M. E. Goddard. 2010. Accurate prediction of genetic values for complex traits by whole-genome resequencing. Genetics 185:623–631.

Meuwissen, T. H. E., Goddard, M. E. (1996). The use of marker haplotypes in animal breeding schemes. Genetic Selection Evolution 28: 161-176.

Meuwissen T, Hayes BJ and Goddard ME.(2001). Prediction of total genetic value using genome-wide dense marker maps. Genetics 157, 1819–1829.

Meuwissen T 2007. Genomic selection: Marker assisted selection on a genome wide scale. Journal of Animal Breeding and Genetics 124, 321–322.

Misztal, I., Legarra, A., and Aguilar, I. (2009). Computing procedures for genetic evaluation including phenotypic, full pedigree, and genomic information. J. Dairy Sci. 92, 4648–4655. doi: 10.3168/jds.2009-2064

Mrode, R., Aliloo, H., Strucken, E. M., Coffey, M., Ojango, J., Mujibi, D., et al. (2018). "The impact of modelling and pooled data on the accuracy of genomic prediction in small holder dairy data," in Proceedings of the World Congress Applied to Livestock Production (Auckland).

Mrode R, Ojango JMK, Okeyo AM and Mwacharo JM (2019). Genomic Selection and Use of Molecular Tools in Breeding Programs for Indigenous and Crossbred Cattle in Developing Countries: Current Status and Future Prospects. Front. Genet. 9:694. doi: 10.3389/fgene.2018.00694

Raj Reshma and Das D. N. (2021). Molecular markers and its application in animal breeding,p. 123-140. Sukanta Mondal and Ram Lakhan Singh (ed.), Advances in Animal Genomics, academic press, ISBN 9780128205952.

Sax, K., 1923. The Association of Size Differences with Seed-Coat Pattern and Pigmentation in Phaseolus Vulgaris. Genetics, 8(6): pp.552- 560.

Silva, R. M. O., Fragomeni, B. O., Lourenco, D. A. L., Magalhães, A. F. B., Irano, N., Carvalheiro, R., et al. (2016). Accuracies of genomic prediction of feed efficiency traits using different prediction and validation methods in an experimental Nelore cattle population. J. Anim. Sci. 94, 3613–3623. doi: 10.2527/jas.2016-0401

VanRaden, P. M. (2008). Efficient methods to compute genomic predictions. J. Dairy Sci. 91, 4414–4423. doi: 10.3168/jds.2007-0980.

Wold, H. 1966. Estimation of principal components and related models by iterative least squares. Multivariate Analysis (Pages 391–420)in P. R. Krishnaiah, ed. Academic Press, New York, NY.

Wold, S., M. Sjostrom, and L. Eriksson. 2001. PLS-regression: A basic tool of chemometrics. Chemom. Intell. Lab. Syst. 58:109–130.

3

Genomic Selection for the Optimization of Breeding Programme: Genomic Selection

Amitosh Kumar[1], Anshuman Kumar[2] and Vineeth MR[3*]

[1]Associate Professor, Department of Animal Genetics and Breeding
[2]Assistant Professor, Department of Animal Genetics and Breeding
[3]Faculty of Veterinary and Animal Science, Rajiv Gandhi South Campus Barkaccha Mirzapur, Banaras Hindu University, Uttar Pradesh

Introduction

The expanding number of people around the world necessitates an increase in the amount of food that is yielded. As stated by the Food and Agriculture Organization of the United Nations (FAO), global food production ought to be ramped up by a factor of two in the next few years in to meet the requirements of the growing global population in terms of consumption (FAO, 2006). This is because the global population is expected to continue growing at a rate of approximately one person every two seconds until the year 2050. It is possible that the predicted production disparity could be filled by making improvements in genetics, animal health, and the excellence of animal husbandry.

The production of animals has expanded significantly over the course of the last several decades as a direct result of the application of quantitative genetics in animal breeding. However, due to technological limitations, the use of genetic markers in breeding programmes has only been implemented to a relatively small extent (Deng et al., 2016). The concept of genomic prediction, which was first introduced approximately twenty years ago, has completely transformed the planning and execution of animal breeding operations.

In today's world, the accessibility of extensive arrays of single nucleotide polymorphism (SNP) markers have presented new prospects for genomic selection, which refers to the process of choosing the animals that would produce the offspring that are genetically superior to the others. Lande and Thompson were the ones who initially proposed the concept of genomic selection (2000). The prediction of breeding value with this technology is accomplished through the introduction of high density SNP genotyping. Utilizing the quantitative trait loci (QTL) that are linked to a certain phenotypic trait in order to exploit them for the goal of selection is the fundamental idea behind this methodology. The extensive information provided by rich genetic markers has made it possible to determine the breeding values of young candidates with greater precision than was previously possible.

According to Konig et al. (2017), genomic selection has resulted in a significant boost in both the economic and technical effectiveness of animal breeding programmes. This effect was initially documented for dairy cattle. These anticipated benefits are typically attributable to a shortening of the time between generations, an improvement in the precision of EBV, and a decrease in the expenses associated with testing progeny (Shumbusho et al., 2015). Following the publication of these underlying theories, genomic selection became widely used in dairy cattle genetic improvement programs

Therefore, the current status of GS in developing countries is presented under the broad subtitles of the stages involved in the implementation of GS such as structure of the reference and validation populations, definition of input variables, genomic prediction models, validation methods, imputation efficiency, genotyping strategies, and routine genomic evaluation. A section on the use of molecular genomic tools and identification of genome-wide signatures of selection is then presented.

Development of Genomic Selection

The application of genomics in breeding programmes was met with immediate success with the development of an innovative new method in which the breeding value may well be determined from markers distributed across the complete genome (de Koning, 2016). Genomic selection is a form of marker-assisted selective breeding (MAS), in which the breeding value of animals can be accurately determined through the use of a rich maker map of chromosomes. However, this method does not provide information about the animals' phenotypes or the phenotypes of their close relatives (Rabier et al., 2016). Genomic selection has led to an increase in animal production by cutting down on the time between generations as well as the expense of high genetic merit bulls (Konig et al., 2009).. To this day, the complete genomes of a wide variety of animals, such as chicken, horse, sheep, cow, dog, cat, and rabbit, have all been sequenced (Fan et al., 2010). The genomic selection is based on the analysis of ranging from ten thousand to eight hundred thousand single nucleotide polymorphisms (SNPs) (Sonesson et al., 2010).

The Breeder's Equation to Achieve Genetic Gain

The utilization of marker data for the purpose of breeding value calculation despite having the knowledge of gene position available is the fundamental premise underlying genomic selection. In order to obtain genotypic information, all of the animals that make up the reference population had their SNPs over the complete genome genotyped (Boichard et al., 2016). Even while the genotyping of large populations comes at a high cost, increasing the number of animals that are part of the reference population would make the results more accurate (Li et al., 2011). The breeder's equation is a monitoring standards that is often used to measure the efficiency of a breeding strategy on genetic gain unit time (Lush 1937). This equation determines the change in a trait that should be anticipated as a result of selection, and it can be estimated using the four main parameters as written.

$$R = i\sigma_a h/L$$

As mentioned by Hickely *et al.* (2017), here R is indeed the variation in trait mean per year or what is known as the response to selection, I would be the selection intensity, σa is the amount of additive genetic variation present in the population, h represents the precision of selection or heritability, and L is the generation interval. Either raising the number of words in the numerator or lowering the generation interval in the denominator is two different ways to achieve the same goal of maximizing response to selection. Genomic selection is an approach that can be utilised to efficiently designate each of these equation's four components.

Candidates having only genotypic information are employed in a prediction equation produced out of a reference population having phenotypes and genotypes. According to Fernandes Junior *et al.*, 2016, it was noted that, bearing the aforementioned considerations in mind, the phonotypic and genotypic data that was obtained were used to build a prediction equation in order to determine GEBV.

The Advantage of Using Genomic Selection

In comparison to more traditional techniques of breeding, genomic selection results in a faster rate of genetic gain, which in turn leads to increased production. In animal breeding systems, the amount of genetic gain, denoted by the symbol (ΔG), is contingent on the intensity of selection, denoted by (i), the accuracy of predictions, denoted by (r), the amount of genetic variance, denoted by (σ2g), and the generation Interval will be denoted by (IG)(Ibañez-Escriche & Gonzalez-Recio, 2011).

$\Delta G = i * r * \sigma^2 g / IG$

The primary advantage of genomic selection would be that individuals can be evaluated even in the absence of information regarding progeny and phenotype. Because of this, screening of animals can begin at a very young age; in fact, it may even be applied to embryos. Genomic selection has the potential to boost production by shortening the time between generations. Additionally, a rise in the size of the reference population may cause an increase in the level of selection intensity.

As a direct result of the development of genomic selection, progeny testing is no longer necessary, which makes the selection process easier to complete and increases its efficiency while reducing the cost of doing so; as an outcome of a significant reduction in the generation interval, the annual genetic trend could be doubled; as a result of their reduced development cost, a much greater number of bulls could well be selected and marketed, which would lead to a better management of genetic resources and limit inbreeding trends (Henryon et al., 2014).

- If marker effect are estimated they can be use for some generation
- Increase genetics gain
- New breeding strategies can be implemented
- Decrease the rate of inbreeding /generation
- Selection of animals before the age of production/reproduction

Whole-genome selection is the most cutting-edge technique for producing and selecting top-notch livestock today.

Utilization of Genomic Selection in Animal Husbandry

Livestock breeders and corporate producers of livestock products can now use gene marker technology to inform their breeding decisions, thanks to advances in gene technology.

The increase in the amount of livestock produced has been greatly aided by advances that have been made in animal breeding. The efficiency of the desired features in animals, such as cattle, sheep, pigs, and poultry, has been improved significantly as a result of significant contributions made through genetic enhancement.

Genomic prediction of production traits in dairy cattle has accuracy more than 0.8, whereas the accuracy for fertility, longevity, somatic cell count, and other traits is 0.7. (Weigel *et al.*, 2012) These exceptionally high degrees of precision are a direct result of the huge reference populations for each breed that have been compiled in order to make genomic prediction possible, as well as the notion that many of the animals that make up reference populations.

Therefore, optimizing genetic gain while limiting the rate of inbreeding will be an important issue for further research as a result of genomic selection. In many nations, heifer calves make up the vast bulk of the animals that are genotyped. Although genotyping early bull calves produces the greatest amount of genetic gain, the cost of doing so is now low enough that it is profitable to genotype heifer calves in order to select which heifers will be kept in the herd (Weigel et al.,2012). The vast majority of selection takes place on the sire side, as bulls' sperm is transferred in artificial insemination (AI), and superior bulls can be chosen at a younger age by means of the application of genomic selection. When the precision of GEBV for a male calf at birth was equal to that of conventional breeding values after the progeny test, this meant that GEBV was a more accurate method. it appeared that cost reductions on the order of 90 percent were attainable (Schaeffer, 2006). When compared to the genotyping of heifer calves, the process of genotyping young bulls not only resulted in an increase in the genetic gain but also resulted in a significant reduction in the cost.

Genomic selection, which can be performed at a fraction of the cost of more conventional breeding techniques, has had a very positive influence on the evolution of dairy cows. Because of the great reliability of genomic prediction derived from low-density genotyping followed by imputation and the relatively cheaper cost of doing so, a very large number of selection individuals have been subjected to the genotyping process. Nearly two million dairy cattle from throughout the world have been genotyped as of this point (Meuwissen et al., 2016).

Future Directions

As a consequence of the likelihood that the need for high-quality proteins, lipids, and micronutrients derived from livestock would rise in tandem with the expansion of the global population, more effective and speedy strategies are required to satisfy the growing requirement for food. Genomic selection presents two opportunities, neither of which have been exploited to their full potential as of yet. If genetic selection combined with reproductive technologies, such as multiple ovulation and embryo transfer, could significantly shorten the time between generations, then this would be a desirable outcome. It is possible to use it to select the embryos that will result in the production of the animals that will come after this one. Furthermore, we may implement the prediction equation on animals that are sold at markets. instead of stud animals, that is, animals that have been measured for the attributes that are important to the commercial sector.

The full genetic data of farm animals may now be accessible extremely readily and at a low price for the purpose of breeding, thanks to advancements in both research and technology. Even while the use of genomic selection is not as well developed in other species of livestock as it is in dairy cow, it is expected that as research continues to advance, the application of genomic selection as a method for the production of superior animals across all species will become widespread. Continued investigation into how genomic selection is implemented in breeding programmes is unquestionably warranted, not just for the sake of the general populace, but also for the benefit of private breeding corporations.

References

Boichard, D., 2016. Genomic selection in domestic animals: Principles, applications and perspectives. Comptes Rendus Biologies, 339(7-8): pp.274-277.

Deng, T., 2016. Four novel polymorphisms of buffalo INSIG2 gene are associated with milk production traits in Chinese buffaloes. Molecular and cellular probes.

Fan, B., 2010. Development and Application of High density SNP Arrays in Genomic Studies of Domestic Animals. Asian Australas. J. Anim. Sci, 23(7): pp.833-847.

FAO (Food and Agriculture Organization of the United Nations), 2006. No World Agriculture: Towards 2030/2050—Interim Report. Food and Agriculture Organization of the United Nations, Rome, Italy.

Fernandes Junior, G.A., 2016. Genomic prediction of breeding values for carcass traits in Nellore cattle. Genetics selection evolution: GSE, 48: p.7.

Henryon, M., Berg, P. & Sørensen, A.C., 2014. Animal breeding schemes using genomic information need breeding plans designed to maximise long term genetic gains. Livestock Science, 166, pp.38- 47

Ibañez-Escriche, N. & Gonzalez-Recio, O., 2011. Review. Promises, pitfalls and challenges of genomic selection in breeding programs. Spanish Journal of Agricultural Research, 9(2): pp.404-413.

König, S., Simianer, H. & Willam, A., 2017. Economic evaluation of genomic breeding programs. Journal of Dairy Science, 92(1): pp.382-391.

Li, H.-D., Bao, Z.-M. & Sun, X.-W., 2011. [Genomic selection and its application]. Yi chuan = Hereditas / Zhongguo yi chuan xue hui bian ji, 33(12): pp.1308- 1316.

Meuwissen, T., Hayes, B. & Goddard, M., 2016. Genomic selection: A paradigm shift in animal breeding. Animal Frontiers, 6(1): p.6

Rabier, C.-E. et al., 2016. On the Accuracy of Genomic Selection. PloS one, 11(6): p.e0156086.

Schaeffer, L.R., 2006. Strategy for applying genomewide selection in dairy cattle. Journal of animal breeding and genetics = Zeitschrift fur Tierzuchtung und Zuchtungsbiologie, 123(4): pp.218-223.

Shumbusho, F., 2015. Economic evaluation of genomic selection in small ruminants: a sheep meat breeding program. Animal, 10(6): pp.1033-1041.

Sonesson, A.K., Meuwissen, T.H.E. & Goddard, M.E., 2010. The use of communal rearing of families and DNA pooling in aquaculture genomic selection schemes. Genetics, selection, evolution: GSE, 42: p.41.

Wiggans, G.R., VanRaden, P.M. & Cooper, T.A., 2011. The genomic evaluation system in the United States: Past, present, future. Journal of Dairy Science, 94(6): pp.3202 -3211.

4

Genomics as Tool for Augmenting Livestock and Poultry Production

Amit Kumar Jha, Rajesh Kumar Vandre, Sulochna Sen S. S. Tomar, A. K. Mishra and N. M. Gupta

Nanaji Deshmukh Veterinary Science University, Jabalpur, Madhya Pradesh

Introduction

Due to global population growth, rising wages, and urbanisation, the demand for animal products has been expanding quickly in many areas of the world (FAO, 2018). Meeting the rising demand for animal products, which is expected to reach 8.6 billion by 2030 (FAO, 2018) as the world's population is projected to increase by 80 million per year (Preisinger, 2012), is one of the biggest difficulties facing modern farm animal industry (Eggen, 2012). Animal products and value-added foods are being consumed as a result of urbanisation and income growth (Hoffmann, 2005; FAO, 2011). By 2050, the world's population will be close to 10 billion, even as the economic conditions continue to improve the demand for animal products will significantly rise. In order to boost animal production, it will be necessary to acquire a more in-depth understanding of animal biology through genomics and other related fields of study. Understanding animal physiology remains a significant problem in the context of sustainable agriculture and animal husbandry, particularly for developing animal farming systems that cater to the new and diversified consumer demand. Livestock and poultry productivity has been greatly increased through traditional selection based on genetic merit estimated from phenotypic and pedigree data. Most breeding plans do not take population impacts on genetic diversity into consideration, and selection is geared more towards the best genetic response for the upcoming generation than the best long-term response (Meuwissen, 1997). Breeders have employed genetic analyses to find superior animals for more than 40 years. The potential of this selection strategy to enhance weakly heritable features without harming production is likewise limited. By focusing on the genetic variations that underlie the phenotypes, genetic progress for yield and other economically significant traits could be advanced even further. However, over the past ten years, academics, producers, and business partners have been very effective in gathering a massive amount of information regarding the genotypes and phenotypes of a huge number of animals. Initially, gaining genotypic information was the main problem. Especially where the economic worth of a single animal is minimal in comparison to genotyping expenses, further work will be required to continue to lower the prices and improve the efficacy of these genotyping platforms. The current challenge, on the other hand, is to predict the phenotype of an animal based on that animal's genotype as well as its environment. This challenge applies to livestock and species that are important to agriculture. Utilizing genomic data to enhance an animal's response to selection is one of the main objectives of animal genomics research. Up until now, phenotypic data have traditionally been used as the basis for genetic enhancement of commercially significant qualities in domestic animals (Boichard et al., 2016). The method of getting phenotypic information for traits that are difficult to measure, low heritable, and tested later in life, however, presents challenges because it is less accurate, expensive, and time-consuming (Goddard and Hayes, 2007). More genetic advancement would result from developments in molecular genetics than from phenotypic information alone because of the potential advantages of employing markers related to genes of interest (information at the DNA level) (Dekkers, 2012; Meuwissen et al., 2016). Modern and advanced genomic tools can be used to quickly progress projects for the genetic enhancement of livestock and poultry. The focus of the interdisciplinary discipline of biology known as genomics is on the structure, function, evolution, mapping, and editing of genomes.

Genomics for Improvement in Livestock

1. Genomic Selection in Dairy Cattle

Dairy cow productivity has been greatly increased through traditional selection based on genetic merit estimated from phenotypic and pedigree data. Theoretically, by focusing on the genetic variations underlying the phenotypes, genetic progress might be expedited even further for yield and other economically significant qualities. Research methods based on genomic technology have been created to find economic trait loci in order to clarify these genetic variations. DNA marker-based assays that identify economic trait loci (ETL) can really be used to improve selection in a business context after the position in the genome has been determined. Breeding values and information from genetic markers that identify desired alleles of economically significant traits could be combined to direct mating choices, leading to genetic improvements over a wider range of attributes. Additionally, the most desired phenotypes influenced by nonadditive gene activity or epistatic interactions between loci could be chosen using marker-assisted selection (MAS).

The first step in creating a commercially effective MAS system is to identify the genetic variation linked to segregating ETL. Resource populations with reasonably accurate phenotypic records and accompanying genomic DNA samples, polymorphic DNA markers that have been placed throughout the genome, a laboratory with the capacity to perform high-throughput genetic analysis of marker samples, and ETL detection algorithms for genetic analysis of genotypic and phenotypic data are the fundamental resources necessary for successfully performing ETL identification across the entire cattle genome. A genetic map with enough relevant markers to maximise genome coverage is one requirement for detecting ETL. Due to the possibility of meiosis and marker heterozygosity limiting map growth and resolution of an ETL, a reference linkage map with a high density of relevant markers ensures flexibility in the choice of markers for genomic scans in experimental populations. Microsatellites make up the majority of the genetic markers on the most recent maps of cattle (Kappes et al., 1997; Barendse et al., 1997). Together, these reference linkage maps indicate the roughly chromosomal locations of more than 1,800 distinctive markers.

2. Genomic Selection in Swine

The selection of elite boar in the nucleus farm, which is followed by the usual boar test records, is the crucial stage in the breeding of pigs. Therefore, generation interval reduction for genetic gain is constrained, but genomic selection may still be able to lower generation interval by up to 25% in comparison to conventional techniques (Meuwissen et al., 2016).

Therefore, by increasing EBV accuracy in pig breeding, genetic gain could be improved, especially for traits that are challenging to improve using conventional methods, such as single-sex, late-in-life, low heritability, and to quantify. Because there are more live piglets than sows are capable of nursing, genomic selection increases piglet pre-weaning mortality even though it increases litter size (Tusell et al., 2013). A broad genomic selection led to a highly accurate selection, which resulted in an increase in the number of teats on sows. This made it possible for sows to provide sufficient care for all of their piglets.

It's noteworthy that using genomic data might potentially boost this trait's reliability by 50%. (Lopes et al., 2017). Another difficult-to-measure attribute is post-weaning mortality, which has a low incidence, poor heredity, substantial environmental impacts, and a high economic value (Knap, 2014). The accuracy of this trait could be improved with conventional techniques, but only in instances where mortality incidence is high. By improving EBV accuracy by up to 50%, the genetic gain of this trait in pigs has also been successfully boosted through the use of genomic selection. (Knol et al., 2016). Cross-breeding is common in pigs, so utilising cross-bred pigs as a reference population would enable for more effective selection.

3. Genomic Selection in Sheep and Goat

Breeds of sheep and goats are chosen for their ability to produce meat, wool, and dairy products, as well as other functional features including reproductive efficiency and disease resistance. Goat and sheep genomes were able to be sequenced from scratch in early 2007 thanks to the development of next-generation sequencing (NGS) (Jiang et al., 2014). It thus presented an opportunity to develop high-density SNP chips. The International Sheep Genomics Consortium created the 54K SNP microarray known as the Illumina OvineSNP50 BeadChip (www.illumina.com) (ISGC; www.sheephapmap.org; Kijas et al., 2009). Similar to this, the International Goat Genome Consortium (IGGC; www.goatgenome.org) was established in 2010 and supported global efforts to construct a 52K SNP microarray for goats that was sold by Illumina. The availability of such high throughput DNA methods and tools in recent years has opened up the use of genome-wide information for sheep and goat breeding. The accuracy gain is closely connected with the reference population size and the genomic heritability of the variable, as demonstrated by Daetwyler et al. (2012), indicating that accuracy and predicted genetic gain may rise in the future if reference populations grow in size.

Additionally, when used on small ruminants, genomic evaluation techniques can significantly increase the GEBV estimation's accuracy, expediting the response to selection. In fact, when the reference population is small, the accuracy of methodologies (such GBLUP and BLUP-SNP) that only consider the phenotypes of the genotyped animals and ignore records from the nongenotyped portion of the population is constrained. Therefore, for such small reference populations, a single-step approach is the suggested technique. It integrates all of the available phenotypic, pedigree, and genomic information in a single-step procedure to calculate genomic breeding values. Additionally, it prevents bias in GEBV estimation brought on by candidate preselection. The method is simple to use since it can use raw phenotypic data without the requirement to compute deregressed proofs, which alter records to account for the fact that animals have varying quantities of information from relatives.

Additionally, it enables the simultaneous evaluation of every animal, genotypes and all. Multi-breed genomic analyses have been favoured due to the variety of meat and dairy sheep and goat breeds and the small population sizes for the majority of those breeds. It is interesting that a significant number of key genes have been found in populations of sheep and goats. Breeders were interested in these key genes because they were linked to a variety of reproductive, illness, or production aspects. Examples of these genes are GDF8 for muscling (Clop et al., 2006), BCO2 for yellow fat, BMP15, GDF9, and FecL genes for hyperprolificacy in sheep etc.

Genomics for Improvement in Poultry

While genomics refers to the collective characterization and quantification of all of an organism's genes, their interrelations, and their influence on the organism, genetics refers to the study of individual genes and their functions in heredity . Genomics involves the sequencing, assembly, and analysis of the entire genome's function and structure. To make even the most complicated biological phenomena easier to grasp, advances in genomics have sparked a revolution in discovery-based research and systems biology. The primary areas of poultry genomics include candidate-gene identification, gene discovery, and the isolation and mapping of genetic markers, which are essential for whole genome mapping and QTL mapping, which involves using marker maps to map QTL (Burt, 2002).

1. Genetic Markers and QTL Detection

In recent decades, there has been a shift away from genetic selection based on phenotypic information and towards genetic selection based on genotype. This change has been brought about by the potential benefits that can be gained from using markers that are linked to genes of interest in breeding programmes. Blood group polymorphisms, proteins, or enzymes were utilised as genetic markers in investigations to find QTL in livestock in the 1950s and 1960s. However, due to the scarcity of genetic markers, these researches were unsuccessful. In order to make this finding, molecular genetics had to advance to the point where it could detect genetic markers in DNA from different animal species. For the purpose of studying the genetic architecture of traits and facilitating the construction of genetic maps, numerous genetic markers have been developed, including restriction fragment length polymorphisms (RFLPs), random amplified polymorphic DNA (RAPD) markers, amplified fragment length polymorphisms (AFLPs), microsatellites, minisatellites, DNA sequence, single strand conformation polymorphisms (SSCPs), and single nucleotide polymorphisms (SNP) etc. By utilizing the association between performance and the transmission of genetic markers in a suitable pedigree, QTLs are identified in the genome. The mapping of the genetic markers that are equally dispersed across the genome is the most crucial step. Microsatellites can be used in this situation because they are widespread, highly polymorphic, and abundant throughout the genome, and because PCR-based tests can be automated for high-throughput genotyping and analysis of big datasets. Successful use of microsatellites to assess the genetic makeup of duck breeds is crucial for the creation of conservation and usage plans for duck genetic resources. In ducks, RAPD markers were discovered to be indicative of biodiversity analyses.

Genomic Selection in Poultry Breeding

As of the 21st century, rapid discoveries in molecular genetics, together with statistical and computational approaches, have made it possible to employ genomic selection (GS) as a new tool to greatly increase genetic gain in animal breeding this was not conceivable before to the 21st century (Stock and Reents, 2013; Heidaritabar, 2016). A type of MAS known as genomic selection, also known as genomic evaluation/prediction or whole-genome selection, uses thousands of SNPs covering the entire genome to predict an animal's GEBV on the assumption that there are many SNPs dispersed throughout the genome and that there are LD relationships between SNPs and QTL (Goddard and Hayes, 2007). GS is a two-step process (Goddard and Hayes, 2007; Wolc et al., 2014; Boichard et al., 2016, Meuwissen et al., 2016). In the initial phase of estimating the effects of a marker or SNP, the reference or training population (i.e., a sizeable population of animals with phenotypes and marker genotypes) is utilised. This is done so that the effects can be accurately estimated. Here, the entire genome's worth of markers (usually more than 10,000 SNPs) is used to genotype individual animals. The selection candidates (also known as the population) with known marker genotypes but unknown phenotypes are subjected to these SNP effects in the second stage. Then, by combining selection candidates' genotypes with the predicted marker effects, the genomic EBV (GEBV) of those candidates is computed. In GS, rather than searching for genes or QTL regions, the only thing that is done is an analysis of the relationships that exist between chromosomal regions of interest in the reference population. These relationships are then used to make predictions about the BVs of individuals who are candidates for selection. The actual execution of this strategy is extremely labor-intensive, both in terms of the number of individuals who need to be genotyped and the number of genome-wide markers that need to be analysed (Boichard et al., 2016). GS has the potential to reduce the number of progenies that need to be phenotyped due to the fact that the elite breeding population (also known as selection candidates) does not necessarily record traits.

The utilisation of high-density SNP in domestic poultry and other domestic species is made possible by high-throughput genotyping techniques. In comparison to dairy cattle breeding, which was the first industry to implement high-density SNP panels, poultry breeding is able to facilitate rapid genetic development thanks to the comparatively large flock sizes and very short generation intervals (Wolc et al., 2016). The development of high-density SNP chips, which make it possible to do genotyping in a quick, widespread, and cost-effective manner, is essential to the viability of GS (Wolc et al., 2016). The research that were conducted on the chicken genome made it possible for many high-density SNPs to become available for GS. These SNPs covered all of the key chromosomes (International Chicken Genome Sequencing Consortium, 2004). Consequently, GS can now be utilised in chicken breeding operations because to this development. As a result of a number of advancements in high-density chip design, the density of SNP panels has increased from 6K to 12K, 42K, 60K, and finally to 600K SNPs. It is anticipated that this will have a considerable effect on the reliability of the GEBV prediction (Wolc et al., 2014; 2016). Numerous chicken breeding companies have made use of this marker information in order to improve their genetic breeding programmes for both layers and broilers. This information has been used by these companies (Fulton, 2012).

Because commercial poultry breeding programmes frequently include sex-limited traits, such as egg production or eggshell quality, as well as many other traits, breeding strategies can significantly benefit from GS through more precise selection at an earlier age and within families. This is due to the fact that GS enables selection to take place within families and at an earlier age (such as disease resistance, feed efficiency, longevity, late production, and production persistency), which are difficult to measure and improve using conventional breeding methods (Stock and Reents, 2013). In the poultry industry, genomic selection is more effective than conventional methods for enhancing low heritability variables (like hatchability and viability). Sex-influenced traits (like bodyweight), and post-slaughtering qualities (like carcass quality). GS offers the chance to perform direct selection for these features on both sexes, early in life and even before hatching (Fulton, 2008; 2012).

As animals under selection can be evaluated and selected without knowledge of their own phenotype or that of their offspring, genomic-based selection can be used at an early age. When compared to conventional selection, genetic selection has the potential to reduce the generation intervals and increase the accuracy of the estimation of BV. This is because genetic selection has the potential to identify selection candidates at an early age and utilise reliable methods of BV prediction. When genomic data were included in addition to the phenotypic information that was available in the conventional selection, an increase in accuracy in the selection process was observed.

References

Andersen, I.L., Naevdal, E. & Boe, K.E., 2011. Maternal investment, sibling competition, and offspring survival with increasing litter size and parity in pigs (Sus scrofa). Behavioral Ecology and Sociobiology. 65(6): p.1159-1167.

Barendse, W., D. Vaiman, S. J. Kemp, Y. Sugimoto, S. M. Armitage, J. L. Williams, H. S. Sun, A. Eggen, M. Agaba, S. A. Aleyasin,M. Band, M. D. Bishop, J. Buitkamp, K. Byrne, F. Collins, L. Cooper, W. Coppettiers, B. Denys, R. D. Drinkwater, K. Easterday,C. Elduque, S. Ennis, G. Erhardt, L. Li, and L. Lil. 1997. A medium-density genetic linkage map of the bovine genome. Mamm. Genome 8:21–28.

Boichard, D., 2016. Genomic selection in domestic animals: Principles, applications and perspectives. Comptes Rendus Biologies, 339(7-8): pp.274-277. Available at: http:// www. sciencedirect.com/science/article/pii/S1631069116300300.

Burt, D. W. (2002). Applications of biotechnology in the poultry industry. World's Poultry Science Journal, 58(1), 5-13.

Clop, A., F. Marcq, H. Takeda, D. Pirottin, X. Tordoir, B. Bibé, J. Bouix et al. 2006. A mutation creating a potential illegitimate microRNA target site in the myostatin gene affects muscularity in sheep. Nat. Genet. 38(7):813–818. doi:10.1038/ng1810

Daetwyler, H.D., A.A. Swan, J.H. van der Werf, and B.J. Hayes. 2012b. Accuracy of pedigree and genomic predictions of carcass and novel meat quality traits in multi-breed sheep data assessed by cross-validation. Genet. Sel. Evol. 44:33. doi:10.1186/1297- 9686-44-33.

Dekkers, J.C.M., 2012. Application of Genomics Tools to Animal Breeding. Current Genomics, 13(3), pp.207-212. Available at: http://www. ncbi.nlm.nih.gov/pmc/articles/ PMC3382275/.

Dong, Y., M. Xie, Y. Jiang, N. Xiao, X. Du, W. Zhang, G. Tosser-Klopp et al. 2013. Sequencing and automated whole-genome optical mapping of the genome of a domestic goat (Capra hircus). Nat. Biotechnol. 31(2):135–141. doi:10.1038/nbt.2478.

Eggen, A. (2012). The development and application of genomic selection as a new breeding paradigm. Animal Frontiers, 2 (1).

FAO. 2011. Biotechnologies for Agricultural Development. Proceedings of the FAO International Technical Conference on "Agricultural Biotechnologies in Developing Countries: Options and Opportunities in Crops, Forestry, Livestock, Fisheries and Agro-industry to face the Challenges of Food Insecurity and Climate Change". Food and Agriculture Organization of the United Nations (FAO). Rome. 569 Pages

FAO. 2018. World Livestock: Transforming the Livestock Sector through the Sustainable Development Goals. Food and Agriculture Organization of the United Nations (FAO). Rome. 222 Pages.

Fulton JE. 2012. Genomic selection for poultry breeding. Animal Frontiers, 2: 30-36. DOI: 10.2527/af.2011-0028.

Goddard ME & Hayes BJ. 2007. Genomic selection. Journal of Animal Breeding and Genetics, 124: 323-330. DOI: 10.1111/j.1439-0388.2007.00702.x

Heidaritabar M. 2016. Genomic selection in egg-laying chickens. PhD Dissertation. Wageningen University, Wageningen, the Netherlands. 220 Pages.

Hoffmann I. 2005. Research and investment in poultry genetic resources – challenges and options for sustainable use. World's Poultry Science Journal, 61: 57-70. DOI: 10.1079/WPS200449.

Jiang, Y., M. Xie, W. Chen, R. Talbot, J.F. Maddox, T. Faraut, C. Wu et al. 2014. The sheep genome illuminates biology of the rumen and lipid metabolism. Science 344(6188):1168–1173. doi:10.1126/science.1252806.

Kappes, S. M., J. W. Keele, R. T. Stone, R. A. McGraw, T. S. Sonstegard,T. P. Smith, N. L. Lopez-Corrales, and C. W. Beattie. 1997. A second-generation linkage map of the bovine genome. Genome Res. 7:235–249

Kijas, J.W., D. Townley, B.P. Dalrymple, M.P. Heaton, J.F. Maddox, A. McGrath, P. Wilson et al. 2009. A genome wide survey of SNP variation reveals the genetic structure of sheep breeds. PLoS ONE 4(3):E4668. doi:10.1371/journal.pone.0004668

Knap, P., 2014. Pig breeding goals in competitive markets. Proceedings, 10th World Congress of Genetics Applied to Livestock Production.

Knol, E.F., Nielsen, B. & Knap, P.W., 2016. Genomic selection in commercial pig breeding. Animal Frontiers, 6(1): p.15.

Lopes, M.S., 2017. Using markers with large effect in genetic and genomic predictions. Journal of animal science, 95(1): pp.59-71.

Meuwissan, T. H. E. 1997. Maximizing the response of selection with a predefined rate of inbreeding. J. Anim. Sci. 75:934–940.

Meuwissen, T., Hayes, B. & Goddard, M., 2016. Genomic selection: A paradigm shift in animal breeding. Animal Frontiers, 6(1): p.6.

Preisinger R. 2012. Genome-wide selection in poultry. Animal Production Science, 52: 121-125. DOI: 10.1071/AN11071

Stock, K. & Reents, R., 2013. Genomic selection: Status in different species and challenges for breeding. Reproduction in Domestic Animals, 48(SUPPL.1): pp.2-10.

Tusell, L., 2013. Genome-enabled methods for predicting litter size in pigs: a comparison. Animal : an international journal of animal bioscience, 7(11): pp.1739-1749.

Wolc, A., 2014. Applications of Genomic Selection in Poultry. Proceedings of the 10th World Congress of Genetics Applied to Livestock Production.

Wolc, A., 2016. Implementation of genomic selection in the poultry industry. Animal Frontiers, 6(February): pp.23-31.

5

Outlining Genome: Wide Association Studies to Reconstruct a Resilient Gene Pool of Indian Livestock Breeds

Deepak Sharma, S.P. Singh and Avneesh Kumar

Department of Animal Genetics & Breeding, College of Veterinary Science & Animal Husbandry, UP Pt Deen Dayal Upadhyay Pashu ChikitsaVigyan Vishwavidhyalaya evam Go Anusandhan Sansthan (DUVASU), Mathura, Uttar Pradesh

The overall number of animals kept as livestock in India comes to 535.78 million, making it the country with the greatest livestock population. According to the 20th Livestock Census, India has a total bovine population of 302.79 Million. This includes Cattle, Buffalo, Mithun, and Yak. There are a total of 109.85 million buffaloes living in the country. The number of goats in the country currently stands at 148.88, while the total number of sheep in the country is 74.26 Million. Most of the breeds in India have evolved naturally through adaptation to varied agro-ecological conditions and have generally been named after their place of origin or on the basis of prominent morphological characteristics. The number of registered breeds in India has increased over time. Majority of livestock breeds are reared by smallholder farmers are of non-descript type. Indiscriminate crossbreeding is extensively occurs resulting in herds of females that may have very complex ancestries.

Breed Purity Determination

Traditionally, the determination of purebred animals is derived from pedigree information. When pedigree information is lacking, alternatively, molecular markers can be used to estimate breed purity. In several livestock species tens of thousands of Single Nucleotide Polymorphisms (SNP) markers located across the whole genome are available. The availability of genotypes of these SNPs allows estimation of breed composition of individual animals using genomic data (Hulsegge et al., 2013). For implementing the such methodology in practice a rapid and reliable method for genetic purity testing of animals is needed, distinguishing crossbred animals from purebred animals and to determine the breed composition. Furthermore, there is genetic variation within breeds and consequently a breed purity test will depend on how well this genetic variation will be reflected in the reference population's dataset. Finally, some introgression of genes of other breeds is generally accepted, e.g. animals registered with 87.5% pedigree purity are generally considered purebred, so the challenge is to determine a threshold value for purity that is generally accepted. Steps of determining the genetic purity of Breeds

(1) Build reference populations with individuals whose breed of origin is known; (2) Select SNP markers that contain sufficient genetic information to be able to discriminate amongst the cattle breeds, (3) Breed assignment and (4) Assignment testing.

A purity test requires genotypes of reference individuals whose breed of origin is known, a so called reference population. The individuals in a reference population should match the full range of genetic diversity within a particular breed. Based on these reference individuals, SNP markers can be selected, which contain sufficient genetic information to be able to discriminate amongst the breeds. The information of the selected SNPs from the reference populations subsequently could be used to infer the ancestry of individuals with unknown origin. For a purity test it is necessary to draw a threshold value for which an allocation of an unknown individual to a breed is accepted.

Genomic Selection (GS)

Breeding aims also include additional functional features like as reproductive performance and disease resistance in addition to the production of meat and milk in livestock. Selection is done in livestock for these production goals. The use of phenotypic, genotypic, and pedigree data in conjunction with genomic selection (GS) paves the way for new opportunities in breeding operations. It could also be beneficial for traits that are assessed later in the life of the breeding females, such as reproductive ability, breeding seasonality, and longevity, in addition to invasive or destructive evaluation measures of carcass composition and meat quality, which are customarily recorded on the relatives of selection candidates and require animals to be sacrificed. For example, reproductive ability is a trait that is measured later in the life of the reproductive females, and longevity is a trait that is measured later in the life of the reproductive females (Daetwyler et al., 2012b).

Next-Generation Sequencing (NGS)

In the beginning of 2007, the advent of next-generation sequencing (NGS) made it possible to de novo sequence the genomes of animals (Dong et al.,

2013). On the other hand, this opened the door for the development of high-density SNP chips. In a same vein, many international consortia have been established in order to promote international cooperation in the creation of a variety of specialized chips for a wide range of species, such as cattle, goats, and other animals (Tosser-Klopp et al., 2014). In the recent years, the accessibility of such high throughput DNA technologies and tools has paved the way for the application of genome-wide information in cattle breeding.

Genome Wide Association Study (GWAS)

On the basis of established methods like genome wide association study (GWAS)can be used to estimate breed composition and purity and to allow incorporating purebred animals in the breed registry for purebred animals. Genotyping and analysing a large number of SNPs is costly and time-consuming. Therefore selecting a subset of SNPs that is sufficiently informative is an important step toward a breed composition and purity test. Several methods can be used to determine which SNPs contain the most information to discriminate between populations. In this study we used the combination of PCA and RF to perform SNP selection. PCA has been used already in cattle to reduce dimensionality of large SNP data sets and to identify breed informative SNPs. When commercially sourced animals are used to generate resource populations to study the genomics of economically important traits such as feed conversion efficiency or diseases, the presence of extensive admixture in the phenotyped and genotyped animals may impact the genome wide association analysis (GWAA) and leads to the training of genomic prediction models in populations for which the breed composition is not understood.

Reference Population

The establishment of a reference population, which must be followed by the execution of dense phenotyping on animals that are genetically similar to the larger population in order to establish a connection between the genotypic information and the phenotypic data, is one of the fundamental aspects that underpin genomic selection. Reference populations found within a country are often made up of a wide variety of breeds as well as crossbreeds. In order to avoid excluding abnormal animals or even entire breeding lines or families, a reference population should accurately represent the variety that is present in a population. This will allow for more accurate analysis of the population. Utilization of breed assignment techniques in a real and practical setting in order to validate the obtained samples' potential utility as members of a reference population. The genomic best linear unbiased prediction (GBLUP) resulted in larger accuracies of EBV than the pedigree-based best linear unbiased prediction (BLUP), but the gain in accuracy was limited for several features and populations. This was the case despite the tiny reference populations. It demonstrated how the gain in accuracy was strongly correlated with both the size of the reference population and the genomic heritability of the trait, which led to the conclusion that accuracy and expected genetic gain are both likely to rise in the future if reference populations are allowed to grow.

Accuracy of Genomic Selection

Being applied to livestock breeds, genomic evaluation approaches have the potential to significantly increase the accuracy of GEBV estimation and, as a result, speed up the response to selection. When the reference population is small, the accuracy of approaches like GBLUP and BLUP-SNP, which use only the phenotypes of the genotyped individuals and ignore the records of the non-genotyped component of the population, is limited. Therefore, a strategy that only requires one step is the one that should be used when dealing with such tiny reference populations. In a single process step, it incorporates all of the phenotypic, pedigree, and genomic insights that is currently accessible, and then calculates genomic breeding values (Misztal et al., 2009). Because candidates are chosen in advance, this method eliminates any possibility of bias in the calculation of GEBV. Because it is able to use crude phenotypic data without having to calculate deregressed proofs, the method is simple to put into practise. Deregressed proofs are a process through which records are modified to reflect the fact that there seem to be varying volumes of data between animals that are inherited from relatives. Additionally, it enables the simultaneous evaluation of all animals, both with and without regard to their genotypes.

Multi-breed genomic assessments have been favoured due to the wide variety of livestock breeds and the relatively small populations of most of those breeds. The advantages of mixing different breeds that had the same breeding goals and were known to have similar characteristics were very variable but generally restricted. In their study on goats, Carillier et al. (2014) compared various different models, including a multi-breed model that combined the two breeds, a per-breed model, and a multi-trait model that considered how each characteristic in one breed was connected to a similar one in the other breed. They discovered that the per-breed model yielded the best regression coefficients. [Citation needed] Accordingly, if one anticipates a large increase in precision for multi-breed genomic analyses, it might be beneficial to use an SNP panel that has a higher density than the existing 50K Beadchip or to impute from sequence data found in critical ancestors. Kneading populations on a global scale might be highly profitable for certain breeds that are bred in multiple countries with the same breeding goals; however, this outcome is contingent on the degree to which populations are genetically connected to one another (through the use of commercial exchanges).

These simulations were based solely on genotyping male selection candidates in order to account for the historically high costs of genotyping. An intriguing approach for maximising the benefits of genomic selection is to use other methodologies, such as combining inexpensive low-density chips or genotype by sequencing with imputation.

Application of Genomic Selection

Characterization of Major Genes

It is worthy of remark that a substantial number of important genes have been isolated from populations of domestic animal species. These main genes were linked to a variety of reproductive, illness, or production characteristics that were of interest to the breeders. These include, for example, the GDF8 gene for muscling (Clop et al., 2006), the BMP15, GDF9, and FecL genes for hyperprolificacy in sheep (Demars et al., 2013; Martin et al., 2014), the Socs2 gene for mastitis susceptibility, and the casein genes for protein content in goat milk (Leroux et al., 1990). A select few were found through research on candidate genes (BMP15 and caseins), but more recently, GWAS with a 50K SNP chip has provided more evidence (Socs2 and Tmem154). It is anticipated that many more significant genes and causal mutations will be available in the not-too-distant future as a result of the reduction in costs associated with sequencing and genotyping, as well as the increase in genomic research conducted on large and small ruminants. Some of those genes are already in use in breeding programmes, such as the -s1 casein gene, which is utilised in French goats. These genes are used, in essence, to pre-select candidates for progeny testing. If there were tiny groups of paternity SNPs or low-density chips with important gene information available, it could be possible to expand this kind of pre-selection method to additional genes and populations. The inclusion of important genes and QTLs with large effects that are connected with a particular trait, on the other hand, should be encouraged in genetic evaluation models in order to reduce the likelihood of bias occurring. All of these factors, taken together, highlight the opportunity for and the pressing necessity of conducting additional research on the genetic and/or genomic evaluation of important genes and significant QTLs in small ruminants. Predicting ungenotyped animals can be difficult, particularly when using a single-step GBLUP technique. Other challenges include modelling multi-allelic variations or QTL haplotypes, as well as merging data from a large number of genes and QTLs. Even though there are opportunities presented by the addition of major genes in genetic and genomic evaluations, it raises the question of how breeding goals should be defined when those major genes have a pleiotropic effect. One example of this is the socs2 mutation, which is unfavourably linked to high milk somatic cell counts as well as being positively associated with greater milk production (Rupp et al., 2015).

Using SNP Chips for Parentage and Assignation

Because of the uneven and restricted application of artificial insemination in most animal breeding programmes, parentage is a contentious subject. In systems which use natural matings using multiple-sire natural mating groups of males for breeding females or in extensive systems, the pedigree information is often unknown. This is also the case in systems that do not use artificial insemination. However, one of the most important aspects that contributes to an increased rate of genetic gain is the pedigree's accuracy and completeness. It has been demonstrated that DNA markers, first microsatellites and more recently SNPs, can be utilised to deduce pedigree information. Using a methodology that is based on the likelihood of events, this set of markers can be put to use not only to identify incorrect parentage but also to select the appropriate parents from among the contenders.

Assessment of Genetic Diversity

The field of genomics presents a wealth of prospects, not only for the genetic development of animals but also as a significant instrument for the evaluation of the genetic variety of native breeds endemic to certain regions. It is possible for SNP chips and whole-genome sequence data to produce estimations of the relationship between animals that are significantly more accurate than those provided by pedigree records (Eynard et al., 2015). This is of utmost significance in circumstances in which complete or accurate pedigree records are either unavailable or lacking. In situations like this, decisions on which creatures should be conserved can be very subjective, which might result in a reduction in genetic variety. This has significant repercussions for conservation efforts both in the wild and in labs (genbanks) (Mucha and Windig 2009). An estimation of common ancestry that is more normally correct can make it possible to make better partner selections, which can help prevent inadvertent inbreeding. In addition, sequence data can enable access to information on rare variants carried by individuals, which is not possible when utilising pedigrees or SNP chips. This information cannot be accessed (which contain mainly the common variants). This is of special interest from the point of view of conservation, with the goal of preserving the greatest amount of genetic diversity possible.

Genomic selection has the ability to improve the robustness of animal husbandry in addition to its other benefits. Genomic selection under these circumstances also has ethical benefits, including a reduction in the number of animals that need to be tested positive for the disease and a reduction in the amount of suffering inflicted on future generations. In a similar vein, research is currently being conducted to enable genetic prediction of characteristics such as feed efficiency and methane emissions. Genomic selection is an apparent option given that measuring these characteristics requires significant financial investment and the use of facilities that are unlikely to be widely implemented on farms (Pickering et al., 2015).

The availability of the genomic sequence in livestock breeds (Dong et al., 2013) has opened up new fields of investigation to better understand and select for traits of interest in this species. The development of a high-density SNP array (Tosser-Klopp et al., 2014) and its application in genome-wide association studies have facilitated the identification of regions that control economic traits.

Conclusion

The utilisation of sequence data has grown rapidly over the past few years, and this trend is only anticipated to continue as technology advances and the price of whole genome sequencing decreases. Whole genome sequence (WGS) data provide access to complete information on all the variants present in an individual, in contrast to Single Nucleotide Polymorphism (SNP) chips, which have recently been used in selection strategies but contain only a subset of the polymorphic variants available in a species. Whole genome sequence data is the sole practical way to obtain the vast majority of the low Minor Allele Frequency (MAF) variations. As a result of taking into account all the genetic diversity, WGS data are anticipated to produce improved estimators of the genuine relationships between individuals. Estimated associations and inbreeding coefficient analysis of the population from which breeding individuals will be selected play a role in breeding decisions. These coefficients can be estimated using either pedigree information, SNP chips, or whole genome sequence data. There has been a noticeable shift in the previous decade from the use of traditional pedigree records in selection tactics to the use of SNP data, which has proven to be effective. Even so, it's possible that using either pedigrees or SNP chips would result in less-than-ideal selections, since pedigrees are sometimes based on incomplete genealogic records and SNP data exhibit ascertainment bias because of the criteria utilised for the chip assembly.

One of the main benefits of WGS is that it allows access to numerous variants with rare alleles that are not captured by SNP chips, in addition to capturing all common variants in the genome. Also, the creation of long-term selection techniques and the influence of short-term vs long-term strategies on livestock species' genetic diversity have received increased focus with the widespread availability of WGS data. This may pave the way for new ways to prioritise animal selection in a conservation-oriented setting and to optimise long-term animal selection. There is a lot of debate over whether or not whole genome sequence data, which is becoming increasingly available, is indeed important for long-term selection techniques and prioritisation of animals to preserve genetic diversity. In the past, researchers have only had access to a pedigree and a few of SNP variants. As a result, many pressing inquiries remain unanswered. When comparing family trees reconstructed using pedigree and SNP data to WGS data (which may include information about rare alleles), how similar or different are the results? If this data is collected, would it be used to develop selection methods that help preserve livestock genetic variety and ensure its continued improvement in the long run? These question can only be answered with availability of more data with the time.

References

20th Livestock Census-2019, All India Report (Available at: https://dahd.nic.in/animal-husbandry-statistics; Accessed 27 December 2021).

Carillier, C., H. Larroque, and C. Robert-Granié. 2014. Comparison of joint versus purebred genomic evaluation in the French multi-breed dairy goat population. Genet. Sel. Evol. 46:67. doi:10.1186/s12711-014-0067-3.

Clop, A., F. Marcq, H. Takeda, D. Pirottin, X. Tordoir, B. Bibé, J. Bouix et al. 2006. A mutation creating a potential illegitimate microRNA target site in the myostatin gene affects muscularity in sheep. Nat. Genet. 38(7):813–818. doi:10.1038/ng1810.

Daetwyler, H.D., A.A. Swan, J.H. van der Werf, and B.J. Hayes. 2012b. Accuracy of pedigree and genomic predictions of carcass and novel meat quality traits in multi-breed sheep data assessed by cross-validation. Genet. Sel. Evol. 44:33. doi:10.1186/1297- 9686-44-33.

Demars, J., S. Fabre, J. Sarry, R. Rossetti, H. Gilbert, L. Persani, G. Tosser-Klopp et al. 2013. Genome-wide association studies identify two novel BMP15 mutations responsible for an atypical hyperprolificacy phenotype in sheep. PLoS Genet. 9(4):E1003482. doi:10.1371/journal.pgen.1003482.

Dong, Y., M. Xie, Y. Jiang, N. Xiao, X. Du, W. Zhang, G. Tosser-Klopp et al. 2013. Sequencing and automated whole-genome optical mapping of the genome of a domestic goat (Capra hircus). Nat. Biotechnol. 31(2):135–141. doi:10.1038/nbt.2478.

Eynard, S.E., J.J. Windig, G. Leroy, R. van Binsbergen, and M.P.L. Calus. 2015. The effect of rare alleles on estimated genomic relationships from whole genome sequence data. BMC Genet. 16:24. doi:10.1186/s12863-015-0185-0.

Hulsegge, B., Calus, M.P.L., Windig, J.J., Hoving-Bolink, A.H., Maurice-van Eijndhoven, M.H.T., Hiemstra, S.J., 2013. Selection of SNP from 50 K and 777 K arrays to predict breed of origin in cattle. J. Animal Sci. 91, 5128–5134.

Leroux, C., P. Martin, M.-F. Mahé, H. Levéziel, and J.-C. Mercier. 1990. Restriction fragment length polymorphism identification of goat alpha s1-casein alleles: A potential tool in selection of individuals carrying alleles associated with a high level protein synthesis. Anim. Genet. 21(4):341–351. doi:10.1111/j.1365-2052.1990.tb01979.x.

Martin, P., J. Raoul, and L. Bodin. 2014. Effects of the FecL major gene in the Lacaune meat sheep population. Genet. Sel. Evol. 46:48. doi:10.1186/1297-9686-46-48.

Misztal, I., A. Legarra, and I. Aguilar. 2009. Computing procedures for genetic evaluation including phenotypic, full pedigree, and genomic information. J. Dairy Sci. 92(9):4648–4655. doi:10.3168/jds.2009-2064.

Mucha, S., and J.J. Windig. 2009. Effects of incomplete pedigree on genetic management of the Dutch Landrace goat. J. Anim. Breed. Genet. 126(3):250–256. doi:10.1111/j.1439-0388.2008.00757.x.

Pickering, N.K., V.H. Oddy, J. Basarab, K. Cammack, B. Hayes, R.S. Hegarty, J. Lassen, J.C. McEwan, S. Miller, C.S. Pinares-Patiño, and Y. de Haas. 2015. Animal board invited review: Genetic possibilities to reduce enteric methane emissions from ruminants. Animal 9(9):1431–1440. doi:10.1017/S1751731115000968.

Rupp, R., P. Senin, J. Sarry, C. Allain, C. Tasca, L. Ligat, D. Portes, F. Woloszyn, O. Bouchez, G. Tabouret, M. Lebastard, C. Caubet, G. Foucras, and G. Tosser- Klopp. 2015. A point Mutation in Suppressor Of Cytokine Signalling 2 (Socs2) increases the Susceptibility to Inflammation of the Mammary Gland while associated with higher Body Weight and Size and higher Milk Production in a Sheep Model. Plos Genetics, in press.

Tosser-Klopp, G., P. Bardou, O. Bouchez, C. Cabau, R. Crooijmans, Y. Dong, C. Donnadieu-Tonon et al. 2014. Design and characterization of a 52K SNP chip for goats. PLoS ONE 9(1):E86227. doi:10.1371/journal.pone.0086227

6

Harnessing Bioinformatics for Sustainable Livestock Production in Genomics Era

Ankit Magotra[1], Kusum Mehla[2]

Department of Animal Genetics and Breeding, Lala Lajpat Rai University of Veterinary and Animal Sciences, Hissar, Haryana, India

Livestock is an integral component of Indian agriculture which is primarily a mixed crop-livestock farming system and plays an important role in the socio- economic life of India. Livestock contributes significantly to the growth and development of agricultural sector by providing a rich source of high quality products such as milk, meat and eggs and by supporting livelihood of more than two third of rural population. Livestock has a crucial and rapidly increasing role infood security and economic stability worldwide Livestock sector contributes for 4.11% of GDP and 25.6% of total Agriculture GDP. The importance of livestock has been undergoing a steady transformation with the contribution of livestock in total agriculture and allied sector Gross Value Added increasing from 24.32% in 2014-15 to 28.63% in 2018-19. Owing to factors such as population and economic growth, urbanization, improvements in transportation and storage practices,its future importanceformeeting human nutritional requirements of food and fiberand a major contributor to agricultural GDP is evident.

To alleviate the concerns raised by a rapidly growing demand of food security and the need of ensuring sustainability of natural resources, we need to better understandthe ways which wouldspeed up commercialization of livestock production so that it can contribute effectively to the bio-economy. Livestock performance and productivity can be improved by adopting proper diet and feeding practices, improved farm management and disease controlstrategies. Selective breeding and genomics play a critical role in enhancement of the genetic potential of different livestock species and reducing the environmental footprint. Since the human genome sequence was first made public in 2001, technical advancements in the fields of molecular biology and genomics have led to an unparalleled rise in the amount of high throughput data that has been collected. The first livestock species to be sequenced was Red Jungle fowl (RJF) chicken. Given the importance of livestock, whole-genome assemblies for important livestock species such as cattle, chicken, pig, sheep, horse, catfish, rainbow trout and other agricultural species have become available within the last 15 years with current efforts invested in improving assembly quality and providing gene annotations.The last decade has witnessed bioinformatics as a key component to analyze huge genomic data and provide optimal solutions. This alignment of genomic data with bioinformatics tools has greatly improved our understanding of genome organization and phenotype decoding which revolutionized genetic and biomedical research.

Crossbreeding of indigenous breeds with exotic species to boost livestock productivity has met with limited success owing to a multitude of factors such as paucityof quality germ-plasm, reduced conception rate and lack of infrastructure. Various whole genome approaches such as Genomic Selection (GS), Genome Wide Association Study (GWAS), and Signatures of Selection (SS) coupled with bioinformatics tools enable large scale study of livestock genomes. This would consequently open up new avenues to deal with future challenge of increased demand for food in a sustainable and environment friendly way. For decades, breeders have tried to enhance the genetic potential of farm animals by altering their genome sequence. They identify markers for desirable phenotypic traits and then select individuals (GS) with such markers to continue their pedigree.GWAS are relatively new which involve examining the genome sequence for variations that might be associated with some particular disease; ultimately underpinning the genes which may contribute to making an individual susceptible to a certain disease. Identification of SS can glean insights into the evolutionary events which shape genomes, functional aspects of genes/genomic regions and genes related to agronomically important traits.

Genomic Selection

GS is a marker-assisted selection method which uses high density markers from across the entire genome, assuring that all Quantitative Trait Loci (QTL) are in Linkage Disequilibrium (LD) with at least one SNP marker. GS works on the principle that information from high density markers dispersed across entire genome can sufficiently capture genome diversity without the need to know exact location of genes. Development of genomic arrays and advancement in genomic technologies have enabled researchers to perform large scale studies in search for mutations underlying economically important traits which would ultimately assist in effective implantation of GS in animal breeding programs.

GS has increased the intensity and accuracy of selection for traits of economic importance, decreased generation interval, increased genetic gain, and allowed detection of recessive lethals. Breast cancer 1 (*BRCA*1) gene was associated with inflammation of mammary gland and production traits in Karan Fries and *CACNA2D*1 gene was associated with mastitis incidence in both Sahiwal and Karan Fries. The accuracy of genomic prediction of genetic values employing computational procedures in dairy cattle is as high as 0.8 for production traits and 0.7 for fertility, longevity, somatic cell count, and other traits.The accuracies of genomic predictions in beef cattle are however lower (0.3-0.7) when compared to dairy cattle. Nevertheless, such high accuracies provide enough motivation to incorporate GS in breeding programs.

Genome Wide Association Study (GWAS)

GWAS is a non-candidate-driven approach as these studies examine the entire genome unlike candidate gene methods which investigate specific genomic regions. GWASs typically focus on examining SNPs across the genome which show strong correlation with a particular disease and have already identified SNPs related to conditions such as diabetes, heart abnormalities, Parkinson disease etc. In livestock, GWAS studies have been successful in identificationof genes associated with agronomically important traits such as beef quality, milk yield and quality (fat and protein percentage), egg production etc. and elucidating the molecular mechanisms underlying these complex traits.

Protein yield, fat yield, and milk yield were all connected with the diacylglycerol O-acyltransferase 1 (*DGAT* 1) gene that is located on the BTA14 chromosome. This gene segregates and affects all milk qualities. Daughters of German Holstein sires were studied. The growth hormone receptor (*GHR*) and prolactin receptor (*PRLR*) genes located on BTA20 chromosome were significantly associated with fat yield and somatic cell score in the population of Holstein-Friesian sires.In chicken, LIM domain-binding factor 2 (*LDB2*) gene on GGA4 chromosome and glypican 6 (*GPC6*) gene had strong association with body growth and weight respectively. The synaptotagmin (*SYT1*), dual adaptor of phosphotyrosine and 3-phosphoinositides (*DAPP1*), protein tyrosine phosphatase, non-receptor type 18 (*PTPN18*) and transcription factor 7-like 2 (*TCF7L2*) are highly significantly associated with feeding behavior traits in pigs.

Selection Signatures

As per classic natural selection theory, mutations that improve the probability of survival of individualshave a propensity to appear more frequently and become fixated in populations over time. Such mutations leave genomic footprints which can be traced using statistical measures. Adaptive evolution of livestock species shaped by artificial selection changed their morphological and behavioral characteristics to an extent that resulted in a wide spectrum of diverse modern breeds.Such features along with comprehensive information about genomic regions associated with agronomically important traits can shed light on genes targeted by selection and provide opportunities to seek insights into the molecular basis of these traits and also assist in developing new methods to detect such selection signatures.

Three QTLs on chromosome BTA6 which affect milk traits in cattle *viz.*ATP-binding cassette, Sub-family G, Member 2 (*ABCG2*); Peroxisome Proliferator-Activated Receptor Gamma, Coactivator 1 Alpha (*PPARGC1A*) and Casein Cluster are under selection in dairy breeds when compared to meat breeds. Increase in allele frequency differences between beef and dairy cattle in the region adjacent to Signal Transducer and Activator of Transcription 1 (*STAT1*) implied that *STAT1*gene implicated in mammary gland development is under selection in Holstein cattle. Insulin-Like Growth Factor 2 (*IGF2*) gene and Melanocortin 4 Receptor (*MC4R*) gene areselected for growth and meat leanness and growth and fatness traits respectively in pigs. In Thoroughbred horses, the Pyruvate Dehydrogenase Kinase, isozyme 4 (*PDK4*) gene, which is related with racing performance characteristics, is currently the subject of selection.

In summary, these genomic resources along with bioinformatics as driving methodology for its versatile capabilities have significant impacts on different aspects of livestock research. Genome sequences are primary source for comprehensive understanding of their gene content, functionality, and networking.Manipulation of genes associated with phenotypic traits of economic value would eventually translate into effective livestock breeding practices leading to improved disease resistance and livestock productivity. The field of agri-genomics through study of livestock genomes enables sustainable productivity for food industry in a climate changing world.

References

Bai Y, Sartor M, Cavalcoli J. Current status and future perspectives for sequencing livestock genomes. J AnimSciBiotechnol. 2012 Mar 1;3(1):8.

deSimoniGouveia JJ, da Silva MV, Paiva SR, de Oliveira SM. Identification of selection signatures in livestock species. Genet Mol Biol. 2014 Jun;37(2):330-42.

Fadiel A, Anidi I, Eichenbaum KD. Farm animal genomics and informatics: an update. Nucleic Acids Res. 2005 Nov 7;33(19):6308-18.

KadarmideenHN.Genomics to systems biology in animal and veterinarysciences: Progress, lessons and opportunities. Livestock Sci. 2014 May;166:232–248.

Magotra A, Gupta ID, Ahmad T, Alex R. Polymorphism in DNA repair gene BRCA1 associated with clinical mastitis and production traits in indigenous dairy cattle. Res Vet Sci. 2020 Dec;133:194-201.

Magotra A, Gupta ID, Verma A, Alex R, Mr V, Ahmad T. Candidate SNP of CACNA2D1 Gene Associated with Clinical Mastitis and Production Traits in Sahiwal (Bostaurusindicus) and Karan Fries (Bostaurustaurus × Bostaurusindicus). AnimBiotechnol. 2019 Jan;30(1):75-81.

Meuwissen TH, Hayes BJ, & Goddard M. (2016). Genomic selection: A paradigm shift in animal breeding. Anim Front. 2016 Jan;6:6-14.

Sharma A, Lee JS, Dang CG, Sudrajad P, Kim HC, Yeon SH, Kang HS, Lee SH. Stories and Challenges of Genome Wide Association Studies in Livestock - A Review. Asian-Australas J Anim Sci. 2015 Oct;28(10):1371-1379.

Womack JE. Advances in livestock genomics: opening the barn door. Genome Res. 2005 Dec;15(12):1699-705.

Yongsheng Bai, Maureen Sartor, and James Cavalcoli. 2012. Current status and future perspectives for sequencing livestock genomes. Journal of Animal Science and Biotechnology. 3:8.

7

Statistical Analysis of Genetic Association Studies Among Unrelated Individuals

Kuldeep Kumar Tyagi

College of Veterinary and Animal Sciences, Sardar Vallabhbhai Patel University of Agriculture and Technology, Meerut -250 110, Uttar Pradesh

People all around the world have been conducting genetic association studies for a long time. Still the literature to practically understand the statistical analysis of such data using software packages is scarce. Statistical analysis of molecular data is a vast topic. So we will be undertaking one of the important basic and preliminary topics *"Statistical analysis of genetic association studies among unrelated individuals"* in this chapter. Overall, we will be discussing population association studies in which qualitative and quantitative data on unrelated individuals for various economic traits is collected and these individuals are also typed at a number of Single Nucleotide Polymorphism (SNP) markers. This excludes family-based association studies, admixture mapping or linkage studies. Various software packages used in this chapter to analyze molecular data of a population are given below in Table 1.

Table 1: Software packages used in this chapter

Software package	Availability
Microsoft Excel	Paid
IBM Statistical Package for Social Sciences Ver. 20.0	Paid
HW_TEST Software	Open Source

Statistical Package for Social Science (SPSS) version 20.0 had been used in this chapter (IBM corp., 2011). Once having well versed with this version a reader can easily understand retrograde as well as forthcoming versions of SPSS easily. This is because there is hardly any change in basic principles of statistics and how they can be used using different versions of SPSS with meagre changes.

1. Coding the Genotypes

To understand practical procedure we will be making use of hypothetical dummy data. This dummy data is available in Excel workbook and can be downloaded online (Download 1). To download the file go to *File=> Download => Microsoft excel.* This data sheet represents SNP data of six genes (Gene 1, Gene 2, Gene 3, Gene 4, Gene 5 and Gene 6) on 550 lactating buffaloes. The associated quantitative data on traits of economic importance are Age at First Calving (AFC) in days, Service Period (SP) in days and Standard 300 day Lactation yield (SLY) in Kg. The qualitative data on mastitis has also been included with numerals 1, 2 and 3 codes for normal, subclinical and clinical mastitis. This excel file can be downloaded from the download option available in File menu.

1.1 Feeding Data in Excel Workbook

First data can be feeded in Excel workbook for the ease of data handling. The header row will define the variables. The first column can be used to either assign the serial number or identification number of the buffalo. Columns thereafter constitute the variables like various genes and traits of economic importance. Data pertaining to each individual is assigned in a single row.

Genotypes are coded for each gene based on the number of alleles identified as shown in Table 2. Say for example serial number 102, the genotypes (coding) were AB(2), AA(1), AB(2), CD(9), AB(2), AB(2) for Gene 1, Gene 2, Gene 3, Gene 4, Gene 5 and Gene 6, respectively. Whereas the values for corresponding traits of economic importance Age at First Calving (AFC) is 1077 days, Service Period (SP) is 121 days and Standard 300 day Lactation yield (SLY) is 2225 Kg, respectively. Additionally this animal is not suffering from mastitis hence categorized under normal (1). Similarly, data pertaining to each buffalo is feeded into the excel workbook. Save your excel workbook on your computer in a folder designated for this analysis.

Table 2: Possible genotypes for different genes and their corresponding coding

Gene 1		A	B			**Gene 2**		A		**Gene 3**		A	B	C
2 Alleles	A	1	2			*1 Allele*	A	1		*3 Alleles*	A	1	2	3
	B		3								B		4	5
											C			6
	AA(1), AB(2), BB(3)						AA(1)				AA(1), AB(2), AC(3), BB(4), BC(5), CC(6)			
Gene 4		A	B	C	D	**Gene 5**		A	B	**Gene 6**		A	B	C
4 Alleles	A	1	2	3	4	*2 Alleles*	A	1	2	*3 Alleles*	A	1	2	3
	B		5	6	7		B		3		B		4	5
	C			8	9						C			6
	D				10									
	AA(1), AB(2), AC(3), AD(4), BB(5), BC(6), BD(7), CC(8), CD(9), DD(10)					AA(1), AB(2), BB(3)					AA(1), AB(2), AC(3), BB(4), BC(5), CC(6)			

2. SPSS Data Sheet

2.1 Importing Data to SPSS

We will be importing data to SPSS once our excel workbook is ready. Open SPSS program installed on your computer. A dialog box as shown in the Figure 1 below will open.

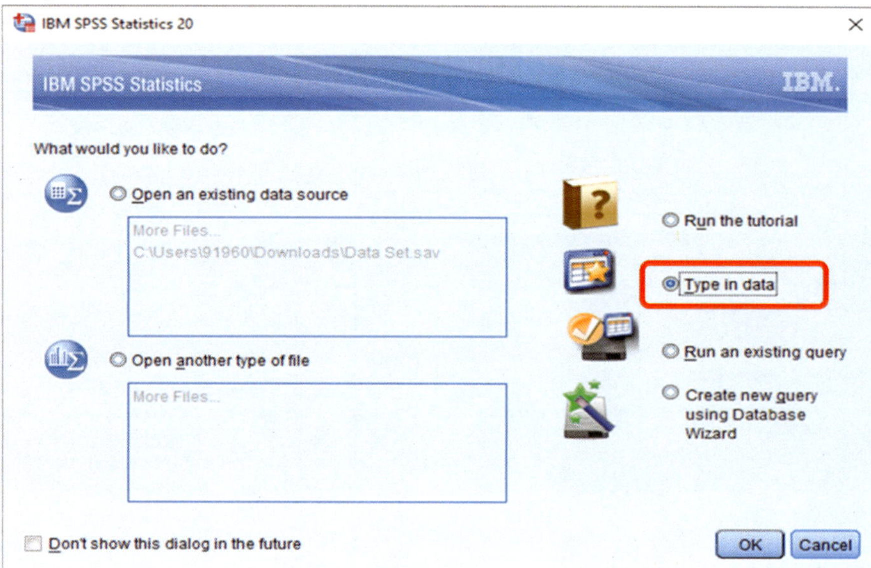

Fig. 1: Dialogue box to open, type data in SPSS

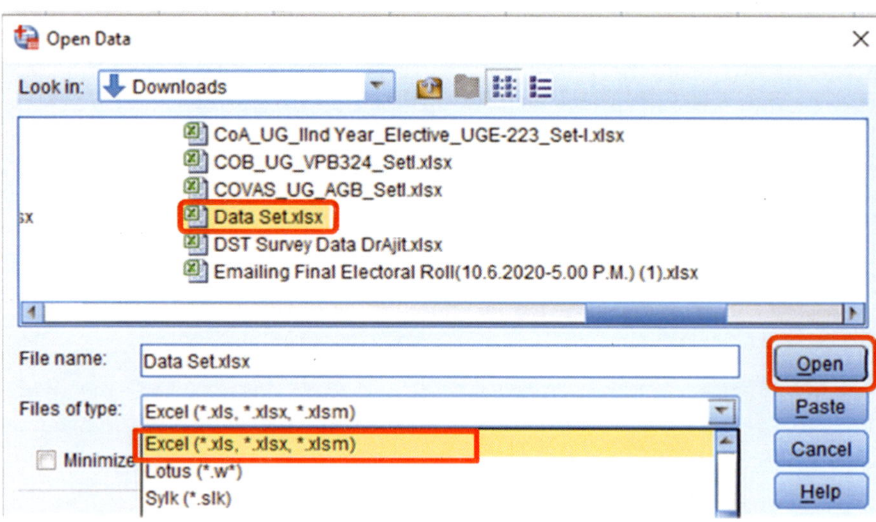

Fig. 2: Dialogue box: Open Data

Click on type in data to open the SPSS data view. Go to *File => Read Text Data*. It will prompt you to locate your data file as shown in Figure 2. Select the File type from the drop down menu and then locate your file and click open. This will open another dialog box as shown in Figure 3.

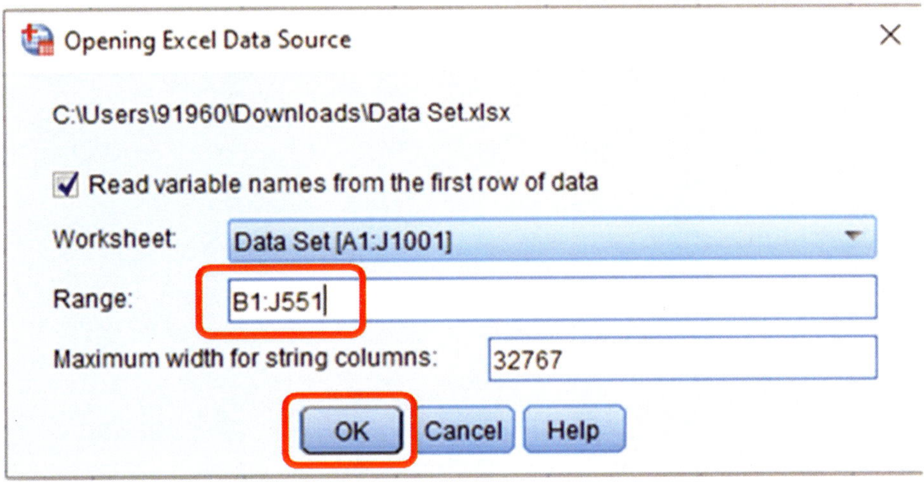

Fig. 3: Dialogue box: Opening Excel Data Source

Define your data range (B1:J551) and then click OK. This will import all your excel data into the SPSS data view. Now we will be defining variables by clicking the variable view tab given at the bottom left of SPSS data view as shown in Figure 4.

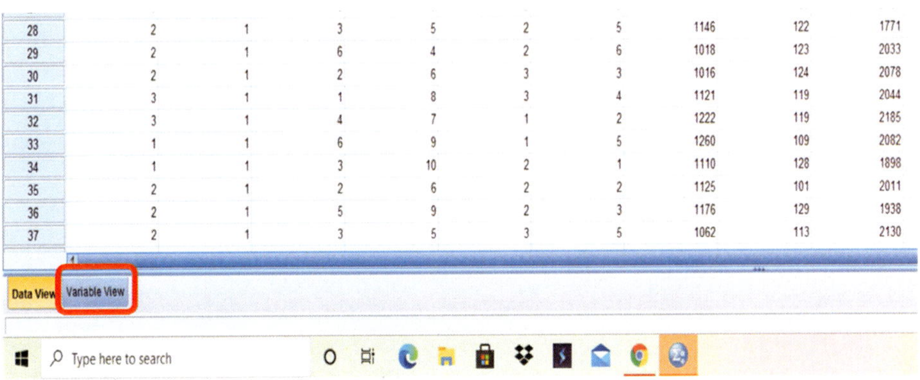

Fig. 4: SPSS Data View

This will open the of the SPSS variable view as shown in Figure 5.

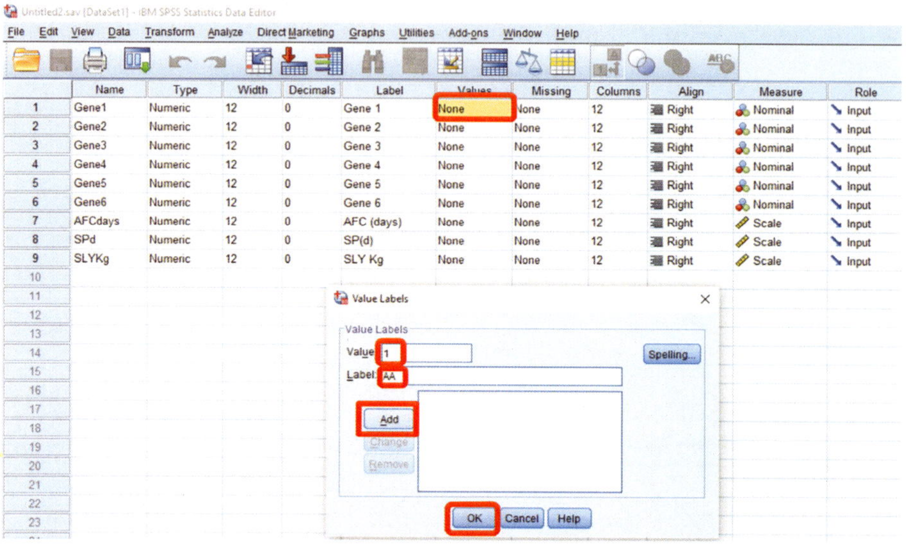

Fig. 5: SPSS Variable view with Value Labels dialogue box

Click in the cell of the column labeled "Values". This will open a dialog box in which value and their associated genotype (Label) can be added. Click on Ok once you complete feeding all the genotype labels corresponding to a gene. The final dataset file can be downloaded online (Download 2).

Now each genotype will be visible in the data view using the toggle button provided at the top menu ribbon in data view of SPSS. So basically if you consider gene 1, which has two alleles (A and B) will be having three genotypes AA, AB and BB and their corresponding coding will be 1, 2 and 3 respectively. Similarly, coding patterns for other genes are also followed in a similar fashion. Coding our data in this way will help us to analyse our data with ease using softwares mentioned in Table 1.

3. Analysis

3.1 Frequency Table Using SPSS

Go to *Analyze => Descriptive Statistics => Frequencies*. This will open a dialog box as shown in Figure 6, select and put all the variable named genes into the variables box and click OK. This will give you an output file with frequency of genotypes for all the genes as shown in Figure 7. The file can be downloaded online (Download 3) for verification of results.

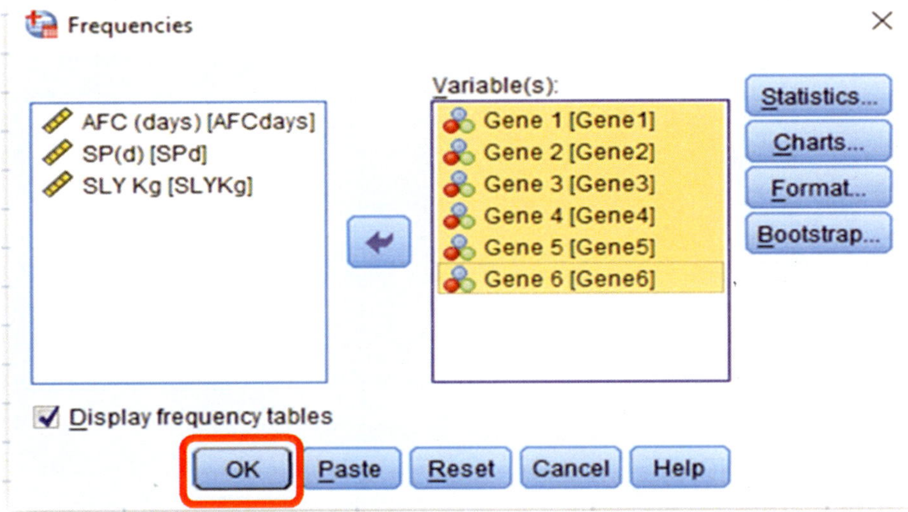

Fig. 6: Dialogue box: Frequencies

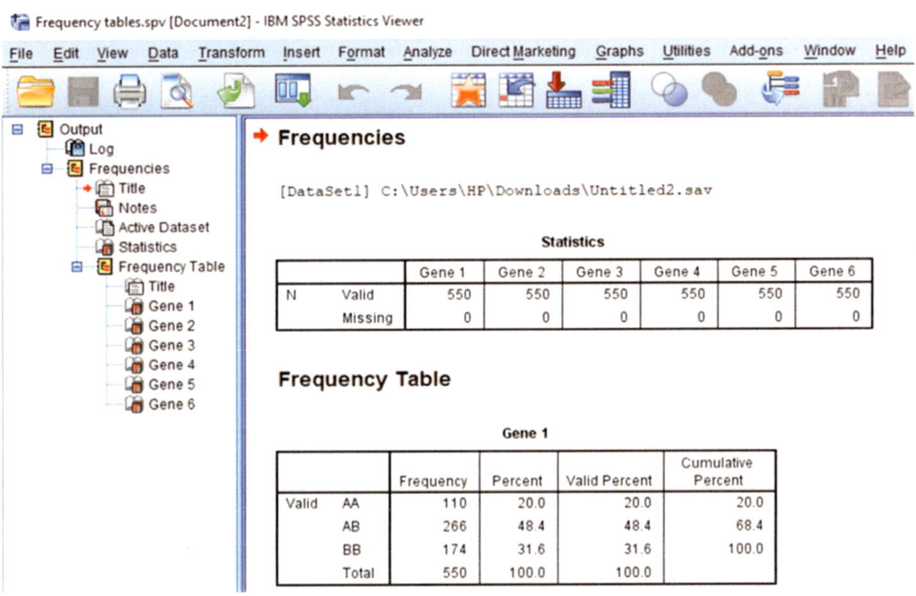

Fig. 7: Output file: Frequency tables

The frequencies so obtained are being feeded in Table 3 and will be used subsequently in Hardy Weinberg equilibrium testing. The benefit of making Table 3 is to have handy absolute frequencies for each genotype ready to be feeded in HW_TEST software. The format prevents mistakes while feeding the value of each genotype because it simulates the data feeding design available in the HW_TEST software.

Table 3: Frequency of various genotypes obtained for different genes

Gene 1		A	B	C	D	**Gene 2**		A	B	**Gene 3**		A	B	C
2 Alleles	A	110	266			*1 Allele*	A	550		*3 Alleles*	A	47	129	115
	B		174								B		77	119
											C			63
Gene 4		A	B	C	D	**Gene 5**		A	B	**Gene 6**		A	B	C
4 Alleles	A	19	84	59	67	*2 Alleles*	A	107	264	*3 Alleles*	A	36	174	177
	B		48	94	75		B		179		B		28	101
	C			29	51						C			34
	D				24									

3.2 Testing for Hardy Weinberg Equilibrium (HWE)

Molecular data on a population are generally subjected to be tested for Hardy Weinberg equilibrium (HWE) as a measure to find preliminary anomalies. HWE test is used extensively on a routine basis to exclude samples with gross molecular typing defects from the usually very large sets of genetic markers presently used in various types of population genetic analyses. Natural or domesticated populations are generally subjected to inbreeding, population stratification, selection or sometimes disease association may lead to deviations from HWE. Sometimes deviation from HWE may be an indication of presence of a common deletion polymorphism, a mutant PCR-primer site or because of a tendency to miscall heterozygotes as homozygotes. So far researchers have tested for HWE primarily as a data quality check and have discarded loci that, for example, deviate from HWE among controls at significance level $\alpha = 10-3$ or 10−4 (Balding, 2006). However, the possibility that a deviation from HWE is due to a deletion polymorphism or a segmental duplication that could be important in disease causation should now be considered before discarding loci.

We will be using user friendly, window based executable open source software HW_TEST for testing HWE (Santos et al., 2020). The program can be obtained free of charge directly from the GitHub internet repository as a standalone executable zip file (Download 4). The downloadable zip file contains a well-defined user manual. The Pearson test is easy to compute, but the $\chi 2$ approximation can be poor when there are low genotype counts, and it is better to use a Fisher exact test, which does not rely on the $\chi 2$ approximation.

3.2.1. Analysis Using HW_TEST Software

The opening window of software HW_TEST asks for the number of alleles in your data set (Figure 8). Let us consider a case of gene 3 which has three alleles. Put this value into the designated field and press next. Now refer to the structural representation of gene 3 and their corresponding genotypes given in Table 3. The values corresponding to number of each genotype need to be filled in designated space provided in dialogue box as shown in Figure 9 of HW_TEST software.

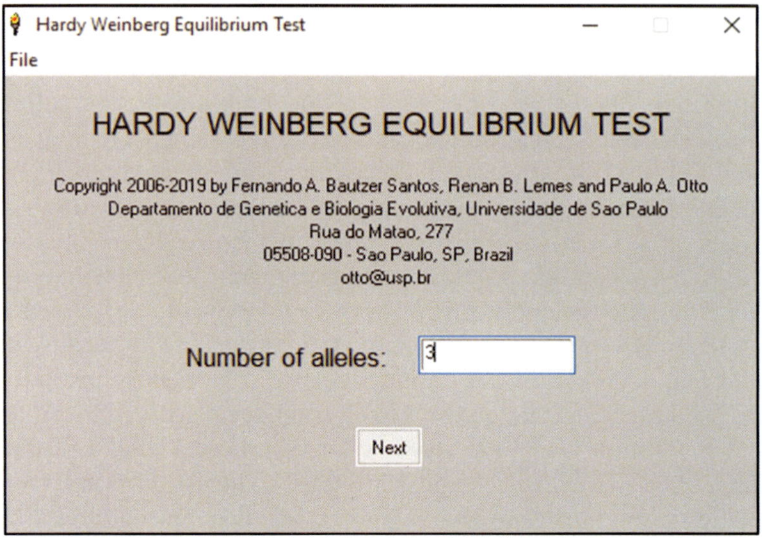

Fig. 8: Opening window of HW_TEST software

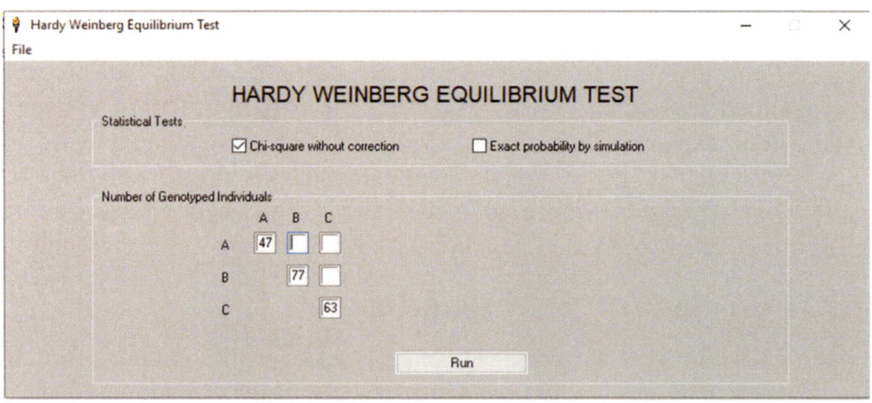

Fig. 9: Dialogue box: Hardy Weinberg Equilibrium Test

Once all the values are entered correctly, click on "Run". This will open the output file with results as shown in Figure 10. The first result output provides us an idea about the observed and expected genotypes frequencies. One can see that the sum of these genotypic frequencies equals unity. Now, the second output provides us with the estimates of allelic frequencies. Finally the P value for estimate is provided. If the value of "P" is greater than 0.05 indicates that our null hypothesis of no difference between observed and expected genotypic frequencies is accepted. Therefore, for the given gene the population is in Hardy Weinberg equilibrium. We can assume for the given gene that, evolutionary forces like mutation, migration and selection are not operating in the given population at the time of screening.

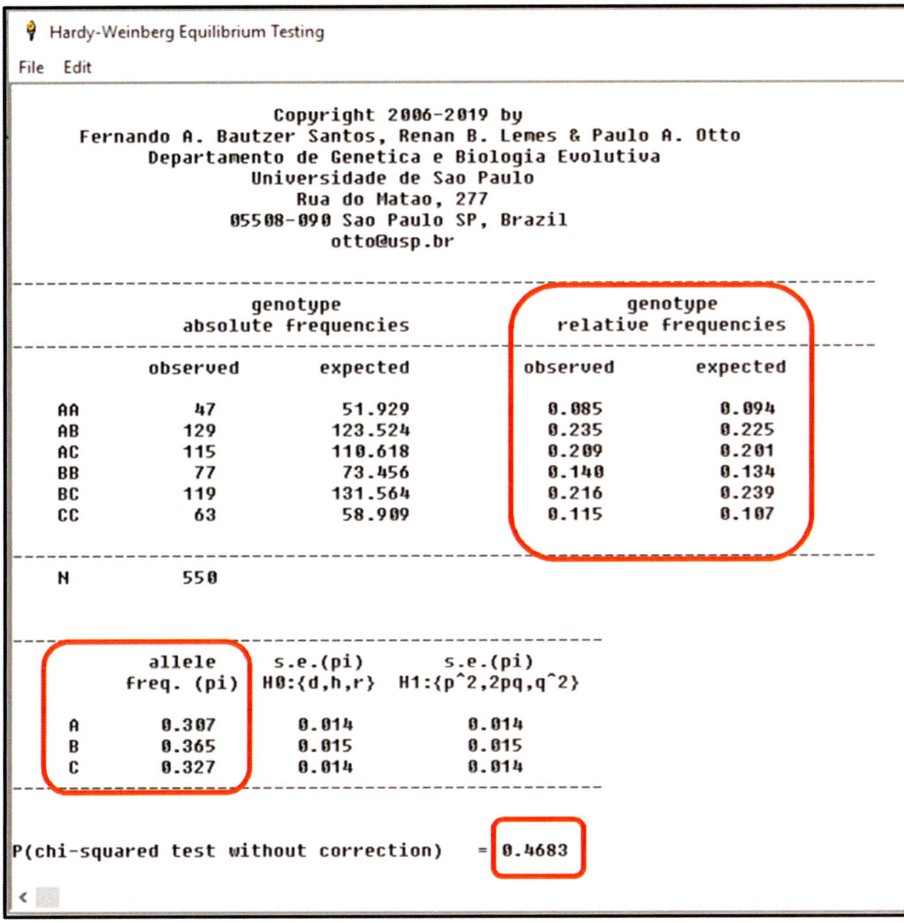

Fig. 10: HW_TEST software result window

Analyzing Association of Genes with Quantitative Traits

Now first we will describe how to analyze association of genes with quantitative data on traits of economic importance like Age at First Calving (AFC) in days, Service Period (SP) in days and Standard 300 day Lactation yield (SLY) in Kg.

Descriptive Statistics using SPSS

To obtain the descriptive statistics we click on Analyze followed by comparing means and then means i.e. *Analyze => Comparing Means => Means*. This will open a dialog box as shown in Figure 11. In the dialog box put all the traits of economic importance into the dependent list and all the genes into the independent list.

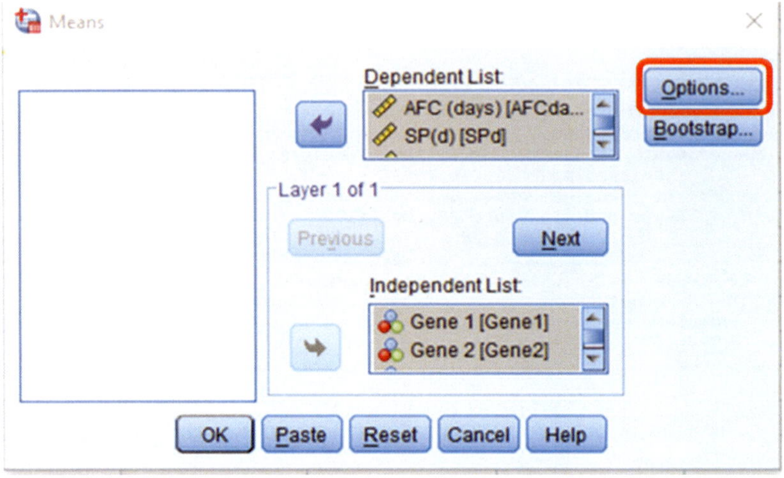

Fig. 11: Dialogue box: Means

Click on the *Options* tab as highlighted in Figure 11. Now this will open another dialog box in which various parameters for descriptive statistics can be defined Figure 12.

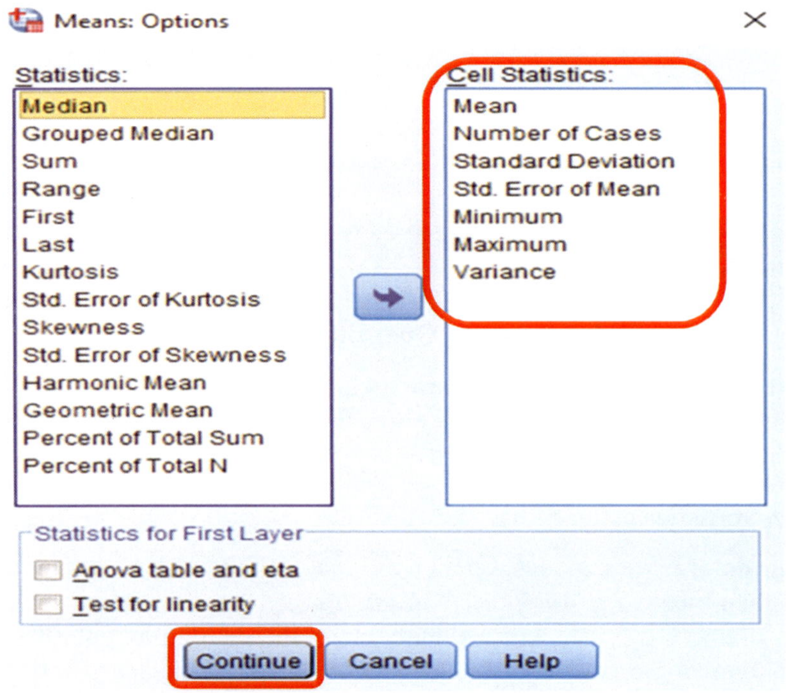

Fig. 12: Dialogue box: Means Options

Thereafter click on *Continue* tab to save the changes and then *Ok* tab to obtain the results of descriptive statistics. The output provides, mean, number of observations, standard deviation, standard error of mean, minimum, maximum and variance pertaining to each genotype Figure 13. The output file can be downloaded online (Download 5) for the ready reference of users to match results during practice of this step.

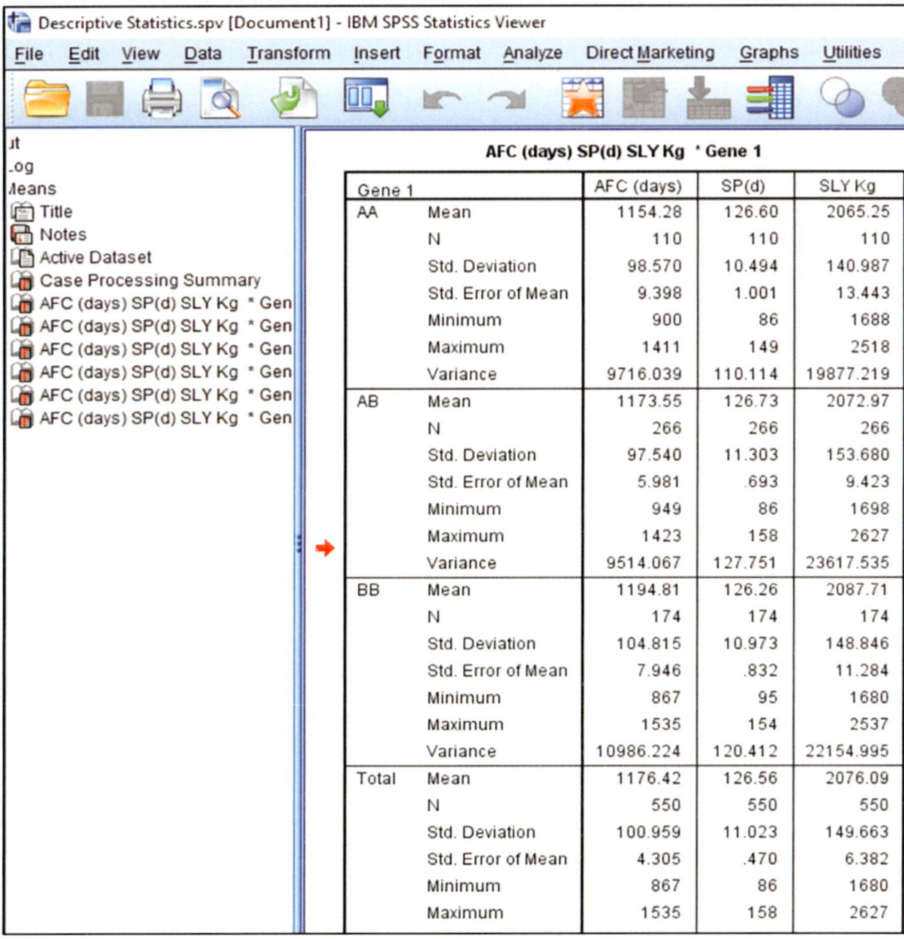

AFC (days) SP(d) SLY Kg * Gene 1

Gene 1		AFC (days)	SP(d)	SLY Kg
AA	Mean	1154.28	126.60	2065.25
	N	110	110	110
	Std. Deviation	98.570	10.494	140.987
	Std. Error of Mean	9.398	1.001	13.443
	Minimum	900	86	1688
	Maximum	1411	149	2518
	Variance	9716.039	110.114	19877.219
AB	Mean	1173.55	126.73	2072.97
	N	266	266	266
	Std. Deviation	97.540	11.303	153.680
	Std. Error of Mean	5.981	.693	9.423
	Minimum	949	86	1698
	Maximum	1423	158	2627
	Variance	9514.067	127.751	23617.535
BB	Mean	1194.81	126.26	2087.71
	N	174	174	174
	Std. Deviation	104.815	10.973	148.846
	Std. Error of Mean	7.946	.832	11.284
	Minimum	867	95	1680
	Maximum	1535	154	2537
	Variance	10986.224	120.412	22154.995
Total	Mean	1176.42	126.56	2076.09
	N	550	550	550
	Std. Deviation	100.959	11.023	149.663
	Std. Error of Mean	4.305	.470	6.382
	Minimum	867	86	1680
	Maximum	1535	158	2627

Fig. 13: Output: Descriptive Statistics

Analysis of Variance (ANOVA) and Duncan test using SPSS

Now we will be subjecting our data for finding any association between various genes studied and traits of economic importance available with us for 550 animals.

Go to *Analyze => General Linear Model => Univariate* from the menu ribbon. This will open another dialog box as shown in Figure 14. Put one quantitative trait (say AFC) into the dependent variable box and all the genes in the fixed factor(s) box. Then click on the *Model* tab, which will open another dialog box to define our model for ANOVA (Figure 15).

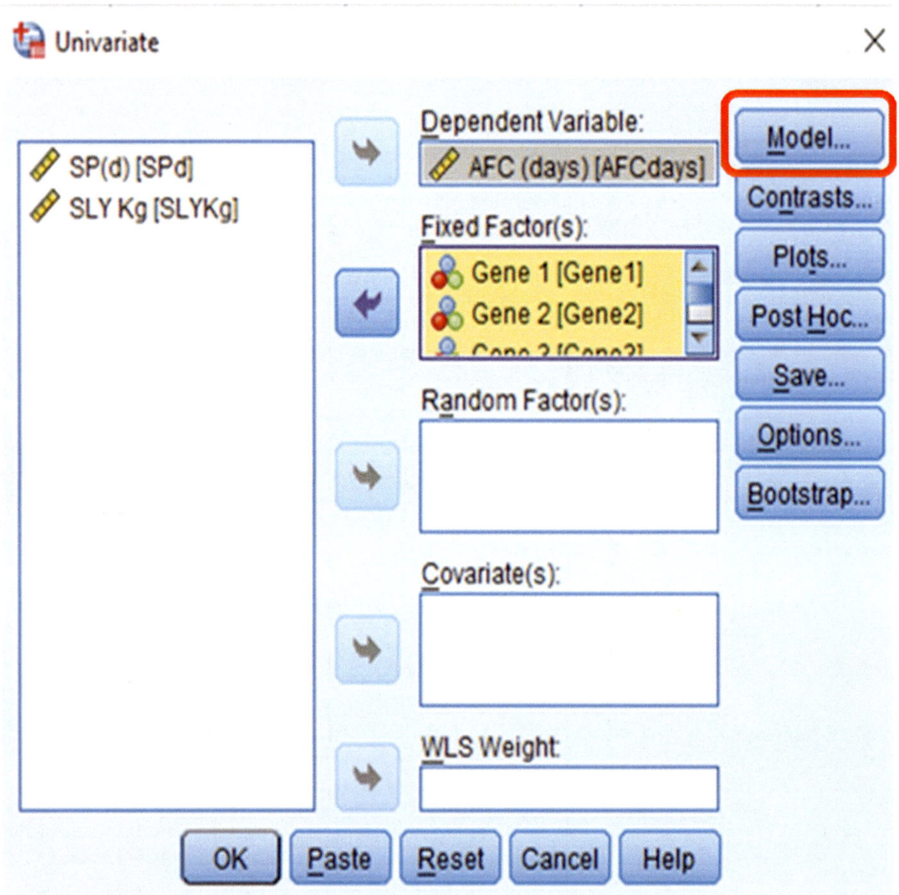

Fig. 14: Dialogue box: Univariate

Click on the custom radio button, select all the fixed factors and choose main effects type from the drop down menu of build terms. Click on the arrow to put all fixed factors into the model box and to activate the continue button. Click on continue to save your changes.

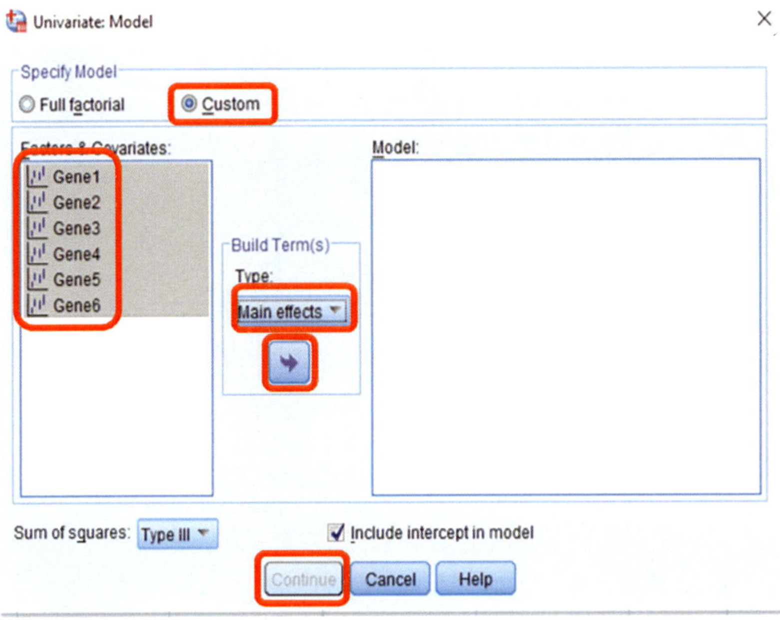

Fig. 15: Dialogue box: Univariate Model

Now click on the *Post Hoc* tab to carry out post hoc tests in your analysis.

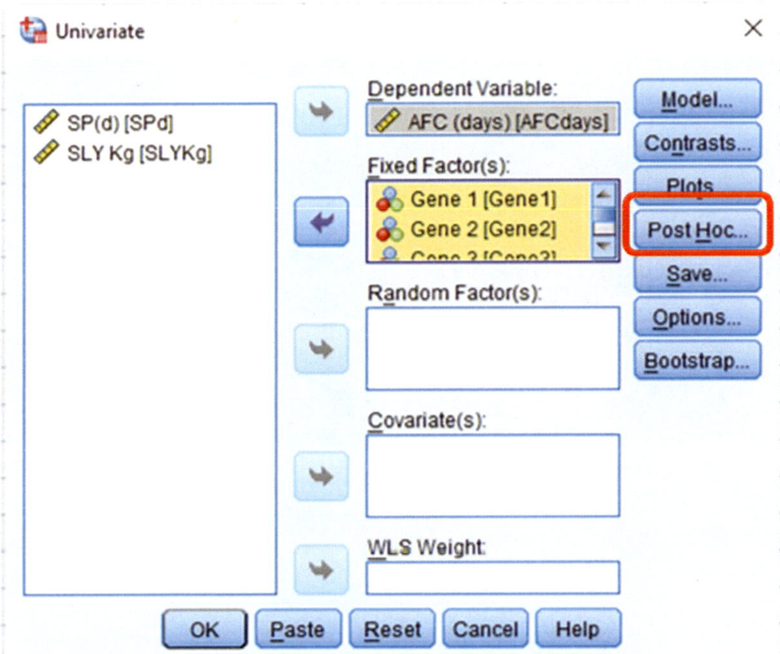

Fig. 16: Activating Post Hoc tests from Univariate dialogue box

This will open another dialog box as shown in Figure 17.

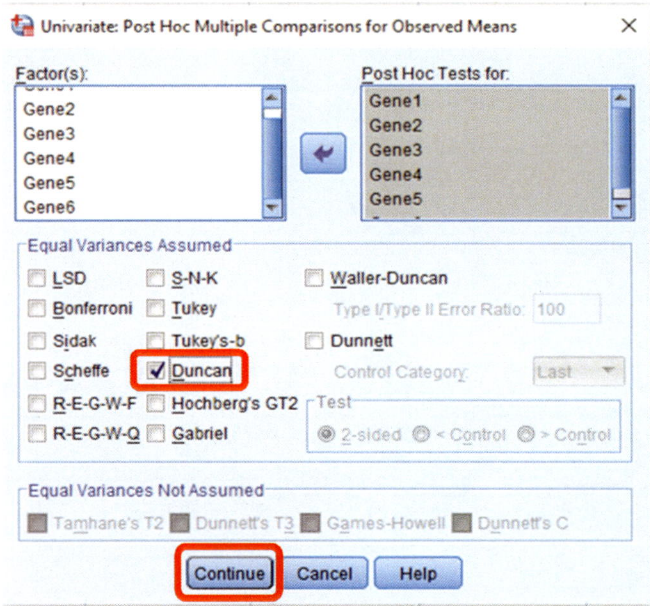

Fig. 17: Dialogue box: Univariate Post Hoc Multiple Comparisons for Observed Means

Put all the fixed factors into the post hoc box, select Duncan checkbox and click on continue to save your changes. Now click on Ok to obtain the results (Figure 18).

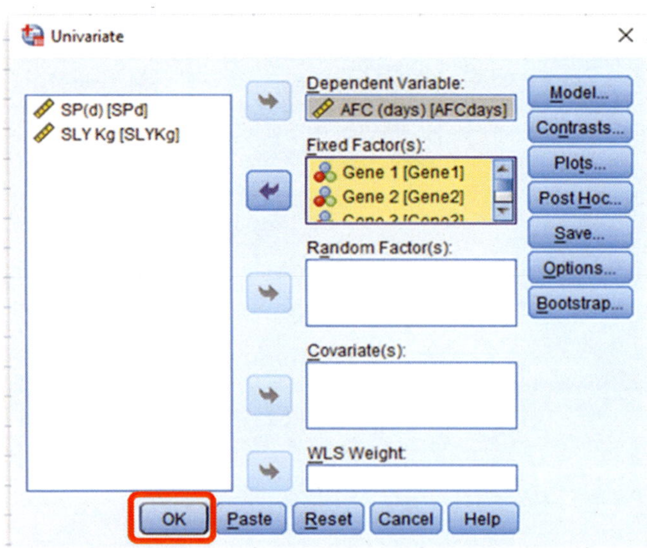

Fig. 18: Executing specified tests from Univariate dialogue box

This will open the output window with desired results. Similarly for other two traits of economic importance viz. SP and SLY we will put them in the dependent variable box one by one and click ok to obtain the results. There is no need to repeat other associated steps as SPSS remember them for the active session before user close down the program. The output file can be downloaded online (Download 6) for ready reference of users to match the results while practicing this step of ANOVA.

The first table we need to look up in the output file is "Test of between subject effects". This is the ANOVA table of our interest (Figure 19).

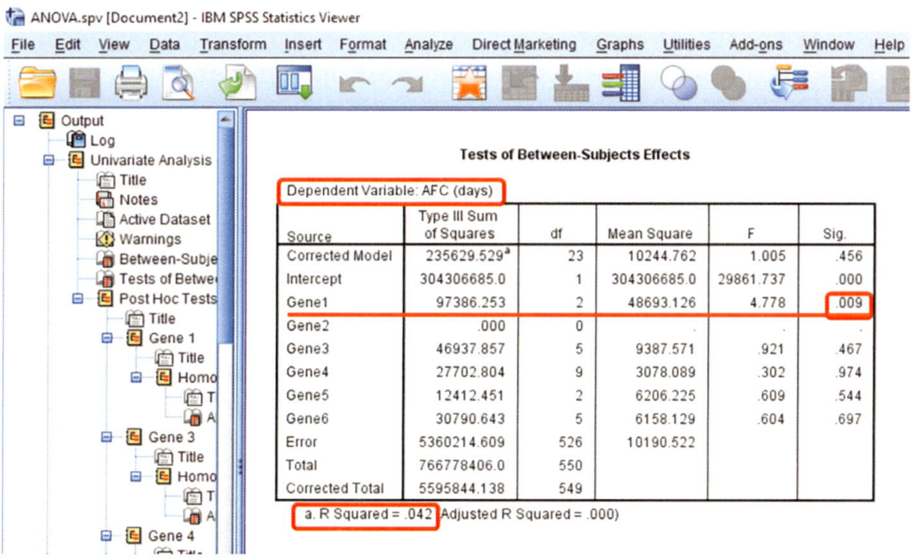

Tests of Between-Subjects Effects

Dependent Variable: AFC (days)

Source	Type III Sum of Squares	df	Mean Square	F	Sig.
Corrected Model	235629.529[a]	23	10244.762	1.005	.456
Intercept	304306685.0	1	304306685.0	29861.737	.000
Gene1	97386.253	2	48693.126	4.778	.009
Gene2	.000	0	.	.	.
Gene3	46937.857	5	9387.571	.921	.467
Gene4	27702.804	9	3078.089	.302	.974
Gene5	12412.451	2	6206.225	.609	.544
Gene6	30790.643	5	6158.129	.604	.697
Error	5360214.609	526	10190.522		
Total	766778406.0	550			
Corrected Total	5595844.138	549			

a. R Squared = .042 (Adjusted R Squared = .000)

Fig. 19: Output window for ANOVA

The table shown in Figure 19 is for the dependent variable AFC. The first column shows the sources of variation which in our case are different genes. Now in this table we have to look for the source of variation which is significantly affecting our dependent variable. The gene 1 in this case is contributing a significant effect with $p = 0.009$ which is less than 0.05. No other gene is causing a significant effect on our dependent variable AFC. The R squared value of 0.042 represents that around 4.2 % variation in our dependent variable is explained by our model. Adjusted R-squared is a modified version of R-squared that has been adjusted for the number of predictors in the model. The adjusted R-squared increases when the new term improves the model more than would be expected by chance. It decreases when a predictor improves the model by less than expected.

Since only gene 1 is contributing significant effect on dependent variable the post-hoc tests pertaining to this gene will only be considered. The output results of post-hoc Duncan test are summarized under the heading of "Homogeneous Subsets" (Figure 20).

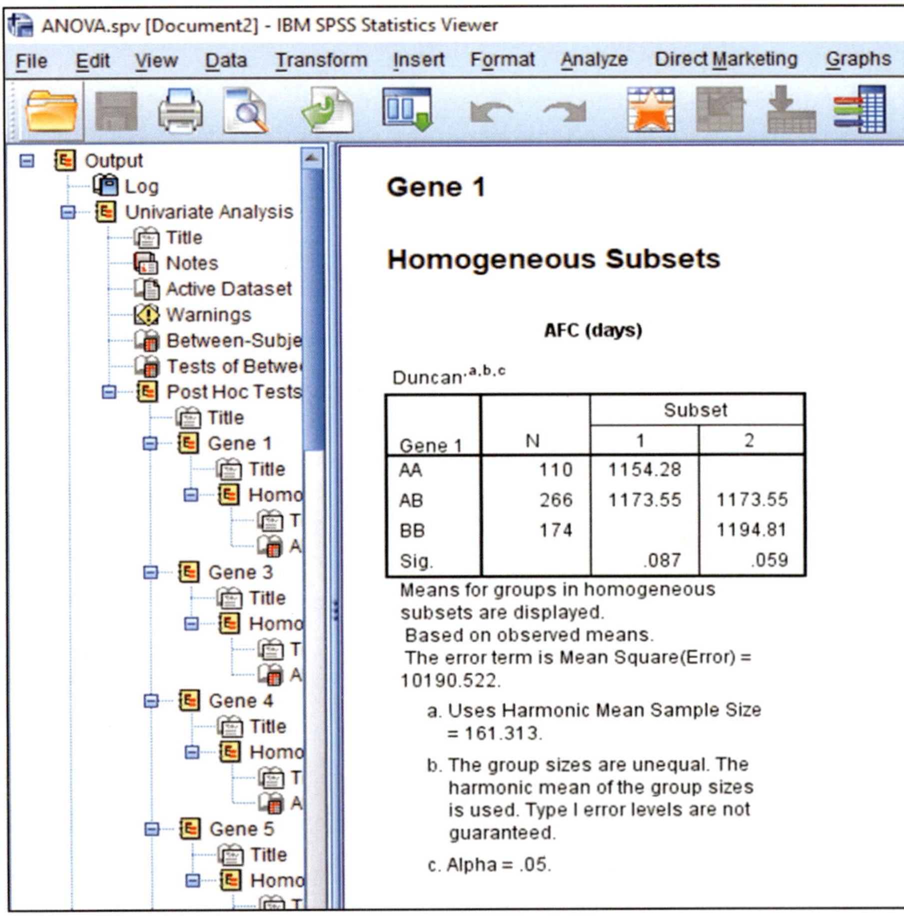

Fig. 20: Homogenous subsets in output file for ANOVA

The output table for Duncan test is quite helpful in determining the superscripts of means that differed significantly ($\alpha = 0.05$). Thus the mean AFC (days) for genotype AA, AB and BB will bear superscripts a, ab and b respectively. The user must note that SPSS provide subsets in the form of numerals (1 & 2 in this case). We generally correspond these subset like 1 $\rightarrow$ a (Read a for 1), 2 $\rightarrow$ b (read b for 2), 3 $\rightarrow$ c (Read c for 3 if it would have been a case in our table, which is of course not there). Now mean for AA genotype resides in subset 1 only, therefore superscript "a" is assigned to it. Genotype AB resides in subset 1 & 2 therefore superscript "ab" is assigned to it. Similarly genotype BB resides in subset 2 only and hence superscript "b" will be assigned to it in table. Similarly, results for all the genes can be reported in similar fashion. This enables us to analyze molecular data for testing Hardy Weinberg equilibrium along the identification of genes affecting dependent variables significantly. The method also gives us an idea about the percent variation explained by our predictor variables in the model.

3.4 Analyzing association of genes with qualitative trait

The animals in the present example have been characterized qualitatively as normal, subclinical and clinical mastitis. The possible significant association of various genes with this qualitative attribute can be analyzed using χ^2Chi square test.

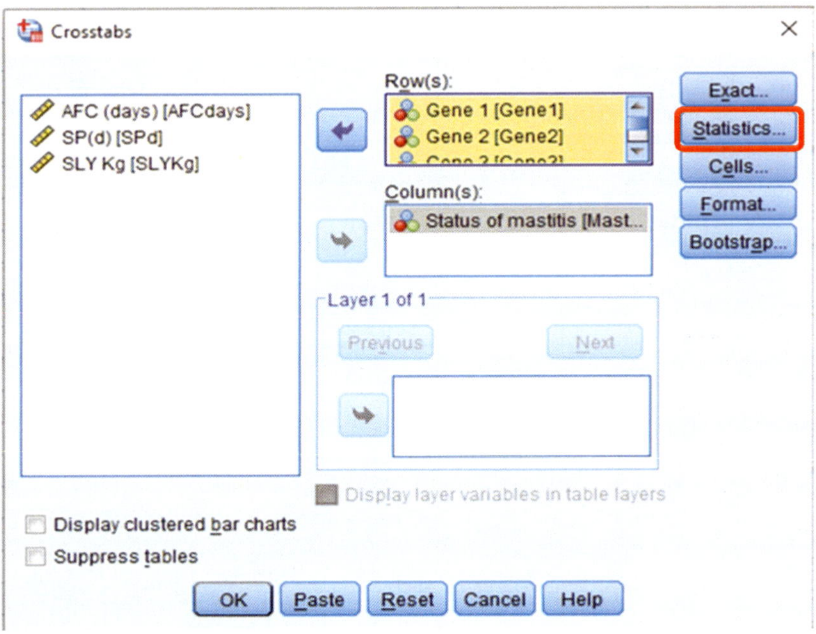

Fig. 21: Dialogue box: Crosstabs

Go to *Analyze => Descriptive Statistics => Crosstabs*. This will open another dialog box Crosstabs (Figure 21). Put the qualitative trait (say status of mastitis) into the column(s) box and all the genes in the row(s) box as shown in Figure 21. Then click on the *Statistics* tab which will open another dialog box to mark tick on Chi square test. Click on continue to save your changes. Now click on the OK button to obtain the results. This will open the output window with desired results. The output file can be downloaded online (Download 7) for your ready reference to match your results while practicing this crosstab chi square analysis.

Gene 6 * Status of mastitis

Crosstab

Count

		Status of mastitis			Total
		Normal	Sub Clinical	Clinical	
Gene 6	AA	8	23	5	36
	AB	43	106	25	174
	AC	46	106	25	177
	BB	3	25	0	28
	BC	24	69	8	101
	CC	10	19	5	34
Total		134	348	68	550

Chi-Square Tests

	Value	df	Asymp. Sig. (2-sided)
Pearson Chi-Square	13.251[a]	10	.210
Likelihood Ratio	17.108	10	.072
Linear-by-Linear Association	.667	1	.414
N of Valid Cases	550		

a. 3 cells (16.7%) have expected count less than 5. The minimum expected count is 3.46.

Gene 4 * Status of mastitis

Crosstab

Count

		Status of mastitis			Total
		Normal	Sub Clinical	Clinical	
Gene 4	AA	4	12	3	19
	AB	25	47	12	84
	AC	11	39	9	59
	AD	10	50	7	67
	BB	10	28	10	48
	BC	21	62	11	94
	BD	29	37	9	75
	CC	10	18	1	29
	CD	13	33	5	51
	DD	1	22	1	24
Total		134	348	68	550

Chi-Square Tests

	Value	df	Asymp. Sig. (2-sided)
Pearson Chi-Square	31.419[a]	18	.026
Likelihood Ratio	33.358	18	.015
Linear-by-Linear Association	1.466	1	.226
N of Valid Cases	550		

a. 4 cells (13.3%) have expected count less than 5. The minimum expected count is 2.35.

Fig. 22: Output results for Crosstab analysis

The first table we need to look up in the output file is Crosstab tables obtained for various genes. These tables depict significant cause-effect relationship between various genotypes of a gene and status of mastitis in concerned animals. If we look at Pearson Chi Square test for gene 6 the P= 0.210, which is greater than the cut off value of 0.05 and hence considered to be non-significant (Figure 22). At the same time when we look at Pearson Chi Square test for gene 4 the P=0.026, which is lesser than 0.05 and hence the effect is considered to be significant. Therefore we can say that various genotypes corresponding to gene 4 have significant effects on the status of mastitis in our case study. The results for remaining genes can also be interpreted in the same way.

4. Summary

This chapter teaches us one of the methods of coding our genotype data. Thereafter data is feeded as each column representing a variable and each row representing observations made on an individual. It has been observed that HW_TEST software is user friendly and proved to be an indispensable tool for testing a population for Hardy Weinberg equilibrium. This software also provides us the estimates of genotypic and allelic frequency in a matter of a few seconds. Further usage of SPSS software has been discussed for finding absolute frequencies of various genotypes pertaining to each gene. SPSS also proved to be user friendly menu driven software to obtain descriptive statistics and ANOVA with post - hoc tests for dependent quantitative variables pertaining to various genotypes. Final interpretation of ANOVA helped us to determine genes causing significant variation in dependent variables. Similarly, SPSS had also been used to find possible effects of various genotypes pertaining to each gene on qualitative variables using crosstab functionality to obtain Chi Square results. Thus, the conjunction of HW_TEST and SPSS software can successfully be used in statistical analysis of genetic association studies among unrelated individuals for various qualitative and quantitative variables.

References

Balding DJ. A tutorial on statistical methods for population association studies. Nat Rev Genet. 2006 Oct;7(10):781-91. doi: 10.1038/nrg1916. PMID: 16983374.

IBM Corp. Released 2011. IBM SPSS Statistics for Windows, Version 20.0. Armonk, NY: IBM Corp.

Santos, F., Lemes, R. B., & Otto, P. A. (2020). HW_TEST, a program for comprehensive HARDY-WEINBERG equilibrium testing. Genetics and molecular biology, 43(2), e20190380. https://doi.org/10.1590/1678-4685-GMB-2019-0380

Supplementary Material Download Links

1. Excel file containing hypothetical dummy data. https://cutt.ly/datafile
2. SPSS Data file. https://cutt.ly/spssdatafile
3. The output file: Frequency tables. https://cutt.ly/frequencytables
4. Open source HW_TEST software. https://cutt.ly/hwetest
5. The output file: Descriptive statistics. https://cutt.ly/means
6. The output file: ANOVA. https://cutt.ly/anova
7. The output file: Chi square. https://cutt.ly/chisquare
8. In case of difficulty in downloading supplementary material online users may reach author at drtyagivet@gmail.com

8

Selection Signatures in Small Ruminant Genome: Concepts and Potential for Improving Performance

Indrajit Ganguly, Nidhi Sukhija, Kanaka K.K., Subrata Koloi Sanjeev Singh and S.P. Dixit

ICAR-National Bureau of Animal Genetic Resources, Karnal, Haryana

Background

Sheep and goats are the poor man's cow because of the lower capital investment, and production costs. Also supplies milk quantities that are suitable for immediate household consumption thereby reducing problems of milk storage and marketing. Small ruminants, goats and sheep, mainly localized in the arid, semi-arid and mountainous areas of the country, have multi-faceted utility in terms of wool, meat, milk, skins, (mo)hair/pelts, manure and have also acquired socio-cultural niches in many communities (Kosgey and Okeyo, 2007). Goat and sheep milk has many advantages over other livestock species' milk being easy in digestion, non-allergic, naturally homogenized, improves immunity as it has antifungal and anti-bacterial properties. However, rearing small ruminants for meat is more popular among farmers as price of chevon and mutton are very high, continuously increasing over the years and sale of male for meat is easy, steady and accessible at farmer's place or local places unlike milk, skin and fibre. Thus, small ruminants (SR) occupy a promising agrarian and economic niche in low to medium input production systems, not suitable for growing crops and dairy farming (Devendra and Chantalakhana, 2002), mostly driven by small and marginal farmers and landless laborers. Their meat and milk production represents 4.8% and 3.4% of the total meat and milk produced, respectively, in the world. Asia contributes maximum production of sheep milk (46%), meat (56%) and wool (48%) compared to the rest of the continents (Figure 1). Also, Asia holds the maximum share in goat milk (60%) and meat (74%) (Figure 2). In India sheep and goat contribute 3% and 5% to meat production respectively and <1% and 3% respectively to milk production (Figure 3). In animal husbandry advanced nations, production systems are well-defined with structured breeding programmes, while in developing countries about 70% to 85% of SR products are based on smallholder and pastoral systems. Small ruminants (goat and sheep) together constitute 223.14 million which is 41.65% of the total livestock population (535.78 million) of India (DAHD, 2019), registering India as the global leader in dairy goats population and goat milk production. The Indian subcontinent houses 37 and 44 well recognized breeds of goats and sheep, respectively (https://nbagr.icar.gov.in/, accessed on 02.02.23). To ensure higher genetic gains from small ruminants, through judicious breeding policies and conservation programmes, and to bridge 'genetic gaps' between agri-demands and agri-supplies (World Bank, 2018), we need to explore their genetic diction to traceback the traits evolutionarily.

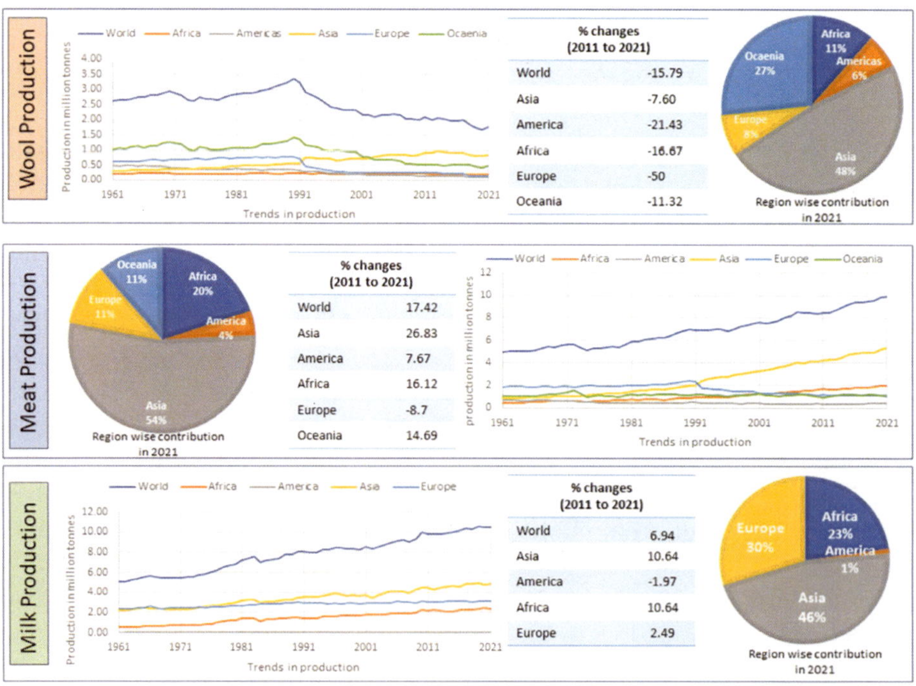

Fig. 1: Status of wool, meat and milk production of sheep in the world (*Source*: www.fao.org)

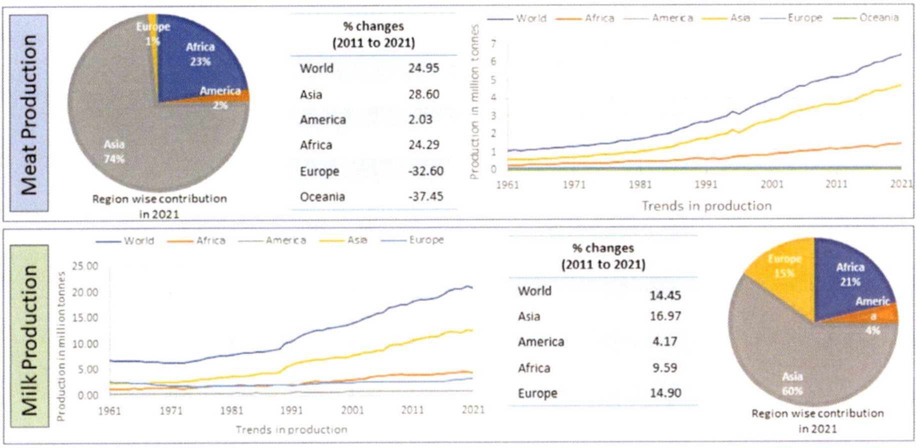

Fig. 2: Status of meat and milk production of goat in the world (*Source:* www.fao.org)

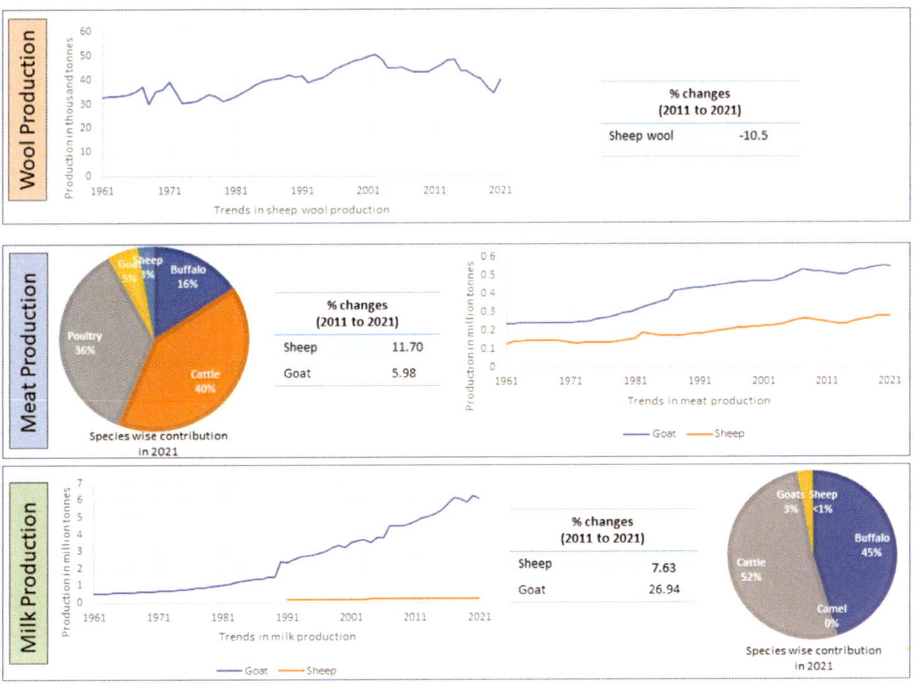

Fig. 3: Status of wool, meat and milk production of sheep and goat in India (*Source*: www.fao.org)

With the development of Next Generation Sequencing (NGS) technologies, it is now possible to precisely characterise whole genome variation in farm animal species to enhance sustainable production. In a population, selection has a particular effect on those genomic areas that are chosen, leaving its genetic imprint. Finding evolutionary traces of the regions that have been under selection since the beginning of domestication in goats 10,500 years ago is largely facilitated by tracing selection signatures. Documenting and validation of selection signatures have a role in taking breeding decisions in a particular agro-climatic region as described in this chapter. Considering its significance, we thoroughly examine SR genetics in this article by recording selection signatures, which reveals variants associated to several traits that can be used to improve small ruminant germplasm.

Selection

Breeders' basic tools for improving livestock are selection and breeding systems. In natural selection, those variations in the genotype (the entire combination of genes inherited from both parents) that increase an organism's chances of reproductive fitness are preserved and multiplied from generation to generation at the expense of less advantageous variations (Falconer, 1996; Templeton, 2009). Positive selection, purifying selection (also known as negative or background selection), and balancing selection are the three types of natural selection. Positive selection favours the novel mutation over the other mutations (Kaplan *et al.*, 1989) due to its selective advantages and the frequency of beneficial alleles increases in the population. The disadvantageous mutant alleles and resulting unfavourable variations are removed in negative selection (Charlesworth *et al.*, 1993), hence the name purifying or background selection. In balancing selection polymorphism is favoured, leading to increased genetic variability. Overdominant selection (in which the heterozygote has a selective advantage), frequency dependent selection (in which various alleles are preferred at different time intervals), and temporally or spatially heterogeneous selection are all examples of this form of selection (Charlesworth, 2006). Each type of natural selection is a consequence to environmental pressure and affects allelic and genotypic frequencies in distinct ways (Oleksyk *et al.*, 2010). Artificial selection, commonly referred to as selective breeding, is a human-mediated process in which qualities/traits (desired by humans) other than fitness (considered by nature) can also be chosen. It is therefore not simply the fitness considerations that determine the gene pool of the next generation. Artificial selection can take two different forms, including methodical and unconscious selection. It has been suggested that the former occurs when there is no long-term goal and is the reason for early domestication. The second occurs when a standard or objective influences how parents are chosen for the next generation (Haldane, 1926; Gouveia *et al.*, 2014). However, the genetic consequences of natural and artificial selection are essentially the same, despite the time frame in which these changes occur is often considerably different (Avise and Ayala, 2009).

Selective Sweep and Selection Signatures

The term "genetic hitchhiking", which is frequently used in population genetics, describes the phenomenon where a population experiencing a selectively favourable mutation is subsequently fixed in that population, affecting the frequency of neutral alleles at surrounding loci (Smith and Haigh, 1974). Each type of selection has unique impacts on selected and neutral loci linked to them (Kreitman, 2000). Berry *et al.* (1991) coined the term "selective sweep," which is now frequently being used to describe genetic hitchhiking caused by directional selection (as suggested by Smith J.M. and Haigh, 1974). Detection of selective sweeps is possible due to three types of distinct signatures that a sweep leaves in genomes.

The *first signature* is the local reduction of the polymorphism levels (Smith J.M. and Haigh, 1974). The *second signature* indicates a particular shift in the site frequency spectrum (SFS) towards low- and high-frequency derived variants (Braverman *et al.*, 1995). Finally, the *third signature* is a localized pattern of linkage disequilibrium (LD) levels, characterized by high LD on each side of a beneficial mutation and low LD between loci that are located on different sides of the beneficial allele (Kim & Nielsen, 2004).

It is also anticipated that the area surrounding the selected site will experience an increase in the average LD resulting into long haplotypes (Kim and Stephan, 2002). This selection signature soon disappears when LD decays and high frequency neutral alleles become fixed in the population following fixation of the chosen mutation (Przeworski, 2002; Kim and Nielsen, 2004; McVean, 2007). Therefore, a recent selection sweep can be indicated by a high frequency of a derived allele and long-range LD (Sabeti *et al.*, 2002; Voight *et al.*, 2006). Additionally, the variability within populations surrounding the selected locus tends to decrease, while between-population diversity tends to increase (Beaumont, 2005; Storz, 2005). Moreover, the number of non synonymous substitutions per non synonymous site (dN) will be higher than the number of synonymous substitutions per synonymous site (dS) (Nei, 2005; Harris and Meyer, 2006).

The selective sweeps can vary in terms of origin, type, and frequency of mutation, and can be classified as hard or soft, complete or partial. A hard selective sweep, also known as a classic selective sweep, occurs when a new allele with a strong advantage rises in frequency quickly until it becomes dominant, reducing genetic diversity in the population. All advantageous alleles in a population can be traced back to a single mutation origin. On the other hand, soft sweeps can either be single-origin or multiple-origin (Hermisson and Pennings, 2017). In the case of single-origin soft sweeps, selection operates on standing genetic variation, which was originally neutral or harmful but has become beneficial or adaptive due to changes in either the environment or the genetic makeup of the population (Hermisson and Pennings, 2005). As the beneficial allele approaches dominance, some genetic diversity remains. In a multiple-origin soft sweep, which occurs in large populations with a higher frequency of mutations, favourable mutations at the same locus occur on different genetic backgrounds, and they all increase in frequency simultaneously, preventing any of them from reaching dominance (Pennings and Hermisson, 2006). In this case, all of the beneficial alleles in a population are frequently found at intermediate frequencies and can be traced back to multiple mutation origins.

In soft sweeps, selection signals are less prominent, and genetic variability in the population is not substantially reduced (Pritchard *et al.*, 2010). As a direct consequence of this, it is far more challenging to identify soft sweeps than it is to detect hard sweeps, and the likelihood of detection is related to the size of the sample. Selective sweeps can be either complete (complete fixation of the preferred allele) or partial (favoured alleles still segregating together with linked neutral sites in the population) (Hermisson and Pennings, 2017).

Need for Selection Signature

Small and scattered herd size threatens to introduce inbreeding in the population (Sharma *et al.*, 2018). Knowledge of genetic diversity of a population is critical for genetic improvement of economic traits and serves as an important guide to update the breeding goals and plans, as per the need and situation (Blott *et al.*, 1988). Complementarily, identification of selection signatures or adapted genotypes may reveal new targets for selection and may lead to better informed breeding decisions. A breed-specific, genome-wide scan for identifying signatures of selection can provide interesting insights into the evolutionary history, adaptation as well as physical and functional information about genes/genomic regions (Mirkena *et al.*, 2010). It can help to highlight new targets for selection and genetic improvement of the breed. For instance altitude adaptation provides insights into a common molecular basis across species and thus a better understanding of the physiological basis of this adaptation. Other significant works related to selected genomic regions for different characters in small ruminants are detailed in upcoming sections. The insights from selection signatures will lead to overall improvement in animal productivity and fitness and will contribute greatly to the economic and food security (Rege and Gibson, 2003).

Identification of Selective Sweeps

The identification of selective sweeps has been addressed through various methods (Table 1). Nicolazzi *et al.* (2015) and Saravanan *et al.* (2020) reviewed the availability of several software suites for this purpose. Hohenlohe *et al.* (2010) developed a decision tree to select selection signatures based on biological factors such as the timescale and mode of selection, demographic fluctuations, population structure, and selected phenotypic traits. These statistics aid in genetic improvement within and between populations. Intra-population statistics and inter-population statistics are the two primary classifications that can be used to describe the methodologies. In the first method, genetic data from within populations are compared to look for signs of selection. This can be done by one of three fundamental methods based on site frequency spectrum, linkage disequilibrium, or reduced local variability (Weigand and Leese, 2018). Inter-population statistics, on the other hand, are primarily based on the degree of difference between populations due to locus-specific allele frequencies and are classified as single-site and haplotype-based differentiation (Zhao *et al.*, 2015). In the following, different methods viz., allele frequency based, LD based methods and composite methods are discussed.

Allele Frequency Based Methods

The skewness in site frequency spectrums, including reduced heterozygosity, an excess of rare variants, an excess of non-ancestral alleles, and valleys of reduced heterozygosity, suggests that selection is taking place. Tajima's D test can detect recent positive selection sweeps by identifying regions with many rare, low-frequency variants caused by recent mutation (Tajima, 1989). In contrast, Fay and Wu statistics evaluate the relationship between ancestral and derived alleles and can detect both positive and negative selection in medium to high-frequency alleles. However, knowledge of ancestral alleles is required for this method (Fay and Wu, 2000). Composite likelihood ratio (CLR) test models the likelihoods of the frequency distribution of being under selection vs neutrality on a chromosome-wise basis, locating selective sweeps and inferring the strength of selection. The higher the CLR value, the more likely it is to be considered a selective sweep (Kim and Stephan, 2002). Recently, the lack of rare (singleton) mutations has been used to detect very recent selection (Field *et al.*, 2016).

Table 1: Commonly used statistics to identify selective sweeps

No.	Approach	Expanded name	Basis	PMID
	D	Tajima's D	Number of segregating sites and the average number of nucleotide differences estimated from pairwise comparison	2513255
	EHH	Extended Haplotype Homozygosity	Detects long range haplotypes (within population)	14764566
	REHH	Relative Extended Haplotype Homozygosity	Compares EHH of tested haplotype against other haplotypes.	25888030
	XPEHH	Across population EHH	Compares EHH of haplotypes near fixation with respect to other haplotypes (between population)	26097641
	Hp	Pooled heterogeneity	Detects reduced local variability using allele counts	20220755
	ROH	Runs of homozygosity	Detects reduced local variability using IBD	18760389
	LK (T_{LK})	Lewontin and Krakauer statistic	Finds if there is significant heterogeneity between loci in their apparent inbreeding coefficients F=s(p)(2)/p(1-p)	4711903
	FLK	Allelic frequency-based population differentiation	Compares observed and expected variances ratio of F_{ST} in population compared to that under neutrality at loci	20855576
	HapFLK	Haplotype-based population differentiation	Uses haplotype information	23307896
	F_{ST}	Fixation index	Population differentiation (between population).	30626638
	CLR	Composite Likelihood Ratio	Compares likelihood of region being selected with neutral model (within population)	29686696
	XP-CLR	Across-Population CLR	Differentiation levels based on CLR (between populations).	26334313
	iHS	Integrated haplotype-homozygosity score	Differential levels of LD around a derived allele compared to ancestral allele (within population)	26314386
	iES	Integrated extended haplotype homozygosity score	By identifying lower levels of EHH decay	28341699
	Rsb	Across population iES	Compares single locus iES (between populations)	27435758

No.	Approach	Expanded name	Basis	PMID
	HMM-SFS	Hidden Morkov Model- Site Frequency Spectrum	Uses SFS and diversity patterns among sequences (within population).	30270233
	CSS	Composite selection signals	Uses rank approach to conclude from F_{ST}, ΔDAF and XP-EHH.	24636660
	d_N/d_S	Non-synonymous substitutions per non-synonymous sites	d_N/d_S=1, >1: positive selection and <1: negative selection.	25388108
	iSAFE	Integrated selection of allele favored by evolution	pinpoint the favored mutation in a large region (□5 Mbp)	29457793
	DCMS	De-correlated composite of multiple signals	Combines different statistics for detecting selection signatures with respect to correlation	25990878

Linkage Disequilibrium (LD) Based Methods

The idea of genetic linkage was first introduced by Morgan in 1911, which explains that genes located near each other on a chromosome are likely to be passed down together during meiotic phase of sexual reproduction. Linkage Disequilibrium (LD) refers to the non-random co-occurrence of alleles at two or more locations, deviating from the expected frequency of random distribution due to independent assortment (Xiong, 2002). LD has the ability to uncover the breeding history of a population and can be the result of admixture, genetic drift (Weir, 2008), or selection and selective sweeps (Sabeti *et al.*, 2002 and Voight *et al.*, 2006). Selection for a favourable allele of a gene increases LD around the selected locus, until recombination reduces the extent of LD. This can be detected by identifying regions of strong LD compared to their prevalence within the population. Alleles at linked locations are referred to as haplotypes. Long range haplotype (LRH), extended haplotype homozygosity (EHH), and relative extended haplotype homozygosity (REHH) are all examples of techniques that can be used to detect selection signatures using LD. Other techniques include integrated haplotype score (iHS) (Voight et al., 2006), nSL (Ferrer-Admetlla et al., 2014), Svd based on haplotype allelic classes(HAC) (Hussin et al., 2010), and ln(Rsb) (Tang et al., 2007) statistics. These methods use extended homozygosity in haplotypes and have the highest power when fixation of the selected allele is reached, while integrated haplotype score has the highest power when selection is ongoing (Ma *et al.*, 2015).

The LRH method measures the homozygosity of haplotypes over a long stretch of genomic regions and aims to uncover the presence of favourable alleles of genes that have been selected for, which results in increased linkage disequilibrium (LD) around the selected locus. The LRH statistic is used to identify loci that have experienced strong positive selection, as the long-range homozygosity of haplotypes surrounding the selected locus is a hallmark of a selective sweep (Sabeti *et al.*, 2002).

Extended Haplotype Homozygosity method detects signatures of positive selection in populations by measuring the extent of homozygosity of haplotypes surrounding a selected locus. It is based on the idea that when a favourable allele is selected for, it increases the homozygosity of surrounding haplotypes, creating a characteristic "decay" of heterozygosity as one moves away from the selected locus. By comparing the EHH of a locus to its surrounding region, one can determine whether it has experienced recent positive selection. The EHH statistic has been widely used in genomic studies to identify loci that have undergone recent positive selection (Sabeti *et al.*, 2002; Voight *et al.*, 2006).

The nSL statistic measures the slope of the distribution of distances between coalescent points and synonymous sites, with the idea that a positive selection event will create a distinctive pattern in this distribution. By comparing the nSL of a locus to that of a surrounding region, one can determine whether it has experienced positive selection (Ferrer-Admetlla *et al.*, 2014).

The HAC-Svd method uses singular value decomposition (Svd) of haplotype frequency data to identify regions of the genome that have undergone positive selection. This is based on the idea that positive selection creates a distinctive pattern in the frequency distribution of haplotypes at a locus, which can be captured by the HAC-Svd analysis. The HAC-Svd method has been used to identify loci that have undergone positive selection in various species, including humans (Hussin *et al.*, 2010).

The ln (Rsb) method uses a log-transformation of the ratio of the average recombination rate (R) to the between-population genetic differentiation (sb) to identify loci that have undergone positive selection. The ln (Rsb) statistic assumes that positive selection causes an increase in the genetic differentiation between populations, and that this increase is proportional to the increase in the recombination rate (Tang *et al.*, 2007).

Differences in Allele Frequency Between Populations

According to Lewontin and Krakauer (1973), neutrality results in low heterogeneity as shown by low F_{ST} values. However, when selection occurs, the within-breed diversity decreases while the between-breed diversity increases, leading to increased variability in F_{ST} (Beaumont, 2005). Wright's fixation index (F_{ST}) is a tool that measures differences in allele frequencies between populations based on individual loci, commonly used in livestock studies to examine population differences. A more advanced method of analyzing population differentiation is the hapFLK metric, which improves upon single locus statistics by considering haplotype differentiation and adjusting frequency estimates based on the Reynolds genetic distances between populations, resulting in breed-specific selection signatures. Paired comparisons are also resistant to demographic effects (Foll and Gaggiotti, 2008).

Composite Method

Composite methods improve the accuracy of test results by combining multiple scores (Fan *et al.*, 2016). The use of multiple approaches enhances the findings by identifying sweeps missed by the previous method and ensuring the reliability of results by reducing false positives caused by population demographic history. Grossman *et al.* (2010) have suggested combining various signals into a composite of signals (termed CMS) mainly to improve the resolution of the detected selection signatures. A keystone of this approach is the ability to simulate data according to calibrated demographic models using the coalescent approach. Driven by the basic concept of the CMS approach, Utsunomiya *et al.* (2013) suggested a new combined statistic, termed meta-SS, merging different genome-wide scan statistics by combining P-values using a method suggested by Whitlock (2005). Randhawa *et al.* (2014) suggested another method termed Composite Selection Signals (CSS) which unifies the multiple pieces of selection evidence from the rank distribution of its diverse constituent tests and was successfully applied to detect selection signatures in cattle and sheep. De-correlated composite of multiple signals' (DCMS) was proposed by Ma et al. (2015) to aggregate several statistics while taking into account their individual correlations.

Selection Signatures Associated with Economic Traits in Sheep and Goat

Recent studies on selection signatures and associated genes related to various economic traits in sheep and goats have been reviewed in this section (Figure 4)

Production

In sheep, selection signatures have been found for wool production traits such as crimp of wool (*PTPN3, NBEA and KRTAP20–2* genes), fiber diameter (*PIK3R4* gene), hair follicle development (*LHX2* gene), growth and development of fiber (*COL17A1* gene) (Mohamadipoor Saadatabadi *et al.*, 2021). *COL17A1* gene has reported to be positively selected for primary follicles formation

(Nie *et al.*, 2018). Zhang *et al.*, 2021 identified selection signatures spanning a set of candidate genes, including *HIF1A*, *CAPN3*, *PRKAA1*, *RXFP2, TRHR* and *HOXA10* that are associated with pathways and GO categories putatively related to hypoxia responses, meat traits and disease resistance in Tibetan and poll desert sheep. *JADE2, PLEKHB2, SPP2, TSHR*, and *UBE2B* genes have been observed to be associated with meat quality traits 3 Indian sheep breed (Saravanan *et al.*, 2021).

For cashmere traits in Chinese goats, potentially associated selective sweeps have been reported in *LHX2, FGF9*, and *WNT2* genes (Wang *et al.*, 2016) as well as *WNT10A* and CSN3 genes (Jin *et al.*, 2021). Candidate genes underlying strong selection signatures including coloration (*ASIP, KITLG, HTT, GNA11*and *OSTM1*) and body size (*TBX15, DGCR8, CDC25A,* and *RDH16*) have also been reported in Chinese goats (Wang *et al.*, 2016). In Ugandan goat genes related to body size and growth (*GJB2* and *GJA3*) and growth and fatty acid composition (e.g., *FGF9* and *IGF1*) have been associated with selection signatures (Onzima *et al.*, 2018). Genome-wide identification of genetic variants and selection signatures have recently been addressed in our laboratory in Changthangi for fibre traits (Vanzampuii, 2022) and in Kanniadu and Jakhrana breeds for meat traits (Ravina, 2022) using whole genome sequence analysis.In Egyptian sheep and goat, body size and development (*BMP2*, *BMP4, GJA3, GJB2*) as well as energy and digestive metabolism (*MYH*, *TRHDE*, *ALDH1A3*) related genes observed to be co-localized with sweeps (Kim *et al.*, 2016). In Dazu black goat and inner Mangolia Cashmere goats selection signatures in "adipocytokine signaling pathway" (2 genes: *IKBKG*, *LOC102190823*), "ether lipid metabolism" (2 genes, *PLD2: PLA2G1B*) and "Glycosphingolipid biosynthesis-ganglio series" (SLC33A1) have been reported (Guan *et al.*, 2016).

Reproduction

In sweeps of Tibetan and poll desert sheep, the thyroid-stimulating hormone receptor gene (TSHR) and ARFRP1 (encodes androgen receptor), which are crucial for prostate gland growth and reproduction, have been identified (Zhang *et al.*, 2021). PPP3CA gene associated with fecundity has been reported in 3 Indian sheep breeds (Saravanan *et al.*, 2021). In Yunling Chinese goat, genes such as *SPEFI*, *CDC25B*, and *MIR103* have reported to be associated with reproduction (Chen *et al.*, 2022). Several candidate genes, including *MARF1*, *SYCP2*, *TMEM200C*, *SF1*, *ADCY1*, and *BMP5*, are involved in goat fecundity in six Chinese goats (Guang-Xin *et al.*, 2019). In Dazu goats, two important and relevant terms, reproductive process and reproduction, were found and included GLRB, AR, PLCB1, NR6A1, and STK3 genes (E *et al.*, 2019). In

Dazu black goat and inner Mangolia Cashmere goats, selection signatures related to "neurohypophyseal hormone activity" (3 genes, *PAIP2B, CCDC64, EPB41L5*), "photoreceptor activity", "spermatid differentiation" (3 genes, *PARD3B, FAT1, KDM4C*), "spermatogenesis" and "male gamete generation" (5 genes, *FAT1, KDM4C, TPPP3, SBF1, PARD3B*) were reported (Guan *et al.*, 2016).

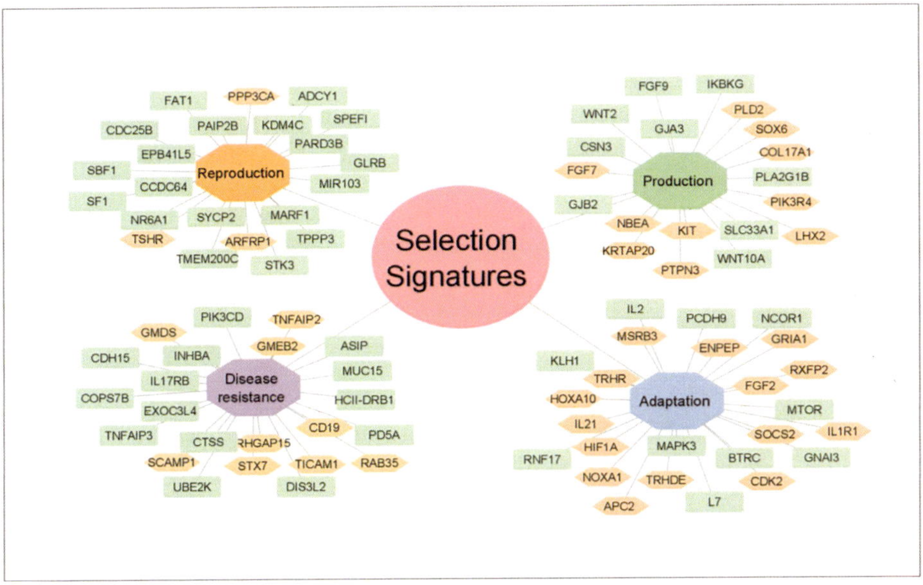

Fig. 4: Network of important genes overlapping with selection signatures associated with different traits in small ruminants. Those mentioned in rectangle are found in goat genome, while those mentioned in hexagon are found in sheep genome

Adaptation

Adaptation is the process by which a species adjusts to changes in its environment, and it is crucial for survival in harsh environments such as high altitudes. In Djallonke and Sahelian sheep three of the genomic regions associated with adaptive selection (*MSRB3* gene, *APC2* gene, and *TRHDE* gene) (Yaro *et al.*, 2019). *TRPM8* gene has been observed to be involved in cold adaptation in 3 Indian sheep breeds (Saravanan *et al.*, 2021). In Tibetan and Poll Dorset sheep, selection signatures spanning a set of candidate genes, including *HIF1A*, *CAPN3*, *PRKAA1*, *RXFP2*, *TRHR* and *HOXA10* that are associated with pathways and GO categories putatively related to hypoxia responses, meat traits and disease resistance (Zhang *et al.*, 2021). Signature associated genes for adaptation to hot arid environments (*CORIN* gene), deficit water status (*CPQ* gene) and heat stress (*PLCB4, FAM107B, NBEA, PIK3C2B* and *USP43* genes) have been identified in Iranian sheep breeds (Mohamadipoor Saadatabadi *et al.*, 2021).

In Central African goat, *PCDH9, KLH1, RNF17, NCOR1*, and *BTRC* have been found in the literature as associated with climatic conditions and in East African goat, signatures targeting *KLH1, PCDH9, RNF17, NCOR1*, and *BTRC* potentially implicated in climate adaptation (Serranito *et al.*, 2021). In Ugandan goats, thermo-tolerance related signatures (e.g., *MTOR* and *MAPK3*) were seen (Onzima *et al.*, 2018). In eight Chinese goat breeds, hypoxia adaptation (*CDK2, SOCS2, NOXA1, and ENPEP*) genes were co-localised with regions of reduced heterozygosity (wang *et al.*, 2016). In Egyptian sheep and goat, thermo-tolerance (melanogenesis) (*FGF2*, *GNAI3*, *PLCB1*), and nervous and autoimmune response (*GRIA1*, *IL2*, *IL7*, *IL21*, *IL1R1*) were identified (Kim *et al.*, 2016). In Mediterranean Sheep and Goats, some of the genes have been implicated in lipid metabolism (*SUCLG2, BMP2*), hypoxia stress/lung function (*BMPR2*), seasonal patterns (*SOX2, DPH6*) or neuronal function (*TRPC4, TRPC6*) (Serranito *et al.*, 2021).

Disease Resistance and Immunity

Selective sweep analysis identified candidate genes associated with disease resistance and immunity, including *GMDS*, *GMEB2*, *TNFAIP3* and *TET2* in Tibetan and poll desert sheep (Zhang *et al.*, 2021). For trypanosomiasis, eight reported genes (*CTSS, ARHGAP15, INHBA, STX7, RAB35, CD19*) were co-localised with regions of reduced heterozygosity in the Djallonke sheep, and the two (*SCAMP1, TICAM1*) were co-localised with regions of reduced heterozygosity in Djallonke and Sahelian sheep. In both the breeds, four regions *(including MHCII-DRB1, PIK3CD, MUC15 and IL17RB* genes) had reduced heterozygosity for *Haemonchus contortus* resistance (Yaro *et al.*, 2019). In Ugandan goat breeds, immune response genes (e.g., *IL10RB* and *IL23A*) were found to be associated with selected regions (Onzima *et al.*, 2018). Whole-Genome Sequencing analysis of Selection Signatures in Hainan Black Goat revealed candidate genes associated with various functions, including immune resistance to disease (*TNFAIP2* (TNF alpha induced protein 2) and *EXOC3L4* (exocyst complex component 3 like 4)), melanin biosynthetic process (*CDH15* (cadherin 15), *ASIP* (agouti signaling protein), and *PARD3* (par-3 family cell polarity regulator)), and light sensitivity (*CNGB3* (cyclic nucleotide gated channel subunit beta 3) and *CNBD1* (cyclic nucleotide binding domain containing 1)) (Chen *et al.* 2022a).In Chinese Yunling Goat, genes such as *DOK2*, *TIMM17A*, *MAVS*, and *DOCK8* related to disease and immunity have found to be co-existed with selective sweeps (Chen *et al.*, 2022b). In African goat, the most plausible signatures of trypanotolerance indicated the genes *DIS3L2*, *COPS7B*, *PD5A*, *UBE2K*, and *UBR1* (Serranito *et al.*, 2021).

Convergent Selection Signatures

Convergent selection is a process of evolutionary convergence where two distinct species or populations evolve similar traits in response to similar ecological pressures (Schluter, 1988; Schaad and Poe, 1988). In sheep and goat populations, convergent selection has led to the development of several similar traits viz., adaptation to grazing, body size and shape, coat color and pattern, Horn development. Since the complexity exists in identifying convergent selection signatures the literature generated in sheep and goats are scanty. Tao *et al.* (2021) by combining genomic and transcriptomic data for the first time identified this genetic convergence in positively selected genes viz., *CHST11* and *SDCCAG8*, in differentially expressed genes viz., *SERPINA14*, *RSAD2*, and *PPIG* at follicular phase, and *IGF1*, *GPRIN3*, *LIPG*, *SLC7A11*, and *CHST15* at luteal phase, and biological pathways including genomic level viz., osteoclast differentiation, *ErbB* signaling pathway, and relaxin signaling pathway, at transcriptomic level viz., the regulation of viral genome replication at follicular phase, and at luteal phase viz., protein kinase B signaling and antigen processing and presentation). Alberto *et al.* (2018) 20 genomic regions were identified as being convergent under selection in Ovis and Capra including *KITLG, HMGI-C, MTMR7, NBEA* etc.,

Potential of Selection Signatures

Selection signatures span knowledge about basic domestication and evolutionary stamps (Nielsen, 2001) to full-fledged functional and ecological characteristics of that region, which are often difficult to trace experimentally. A breed-specific, genome-wide scan for identifying signatures of selection can provide some interesting insights into the evolutionary history, adaptation as well as physical and functional information about genes/genomic regions (Mirkena *et al.*, 2010). They can identify the beneficial polymorphisms in a population (Schlötterer, 2003). They can be used to validate quantitative trait loci (QTL) mapping experiments by reporting causal mutations in regions previously identified by QTL mapping experiments (Gouveia *et al.*, 2014). Knowledge of genetic diversity of a population is critical for genetic improvement of economic traits and helps to update the breeding goals, as per the need and situation (Blott *et al.*, 1988). For instance, small and scattered herd size invites inbreeding in the population (Sharma *et al.*, 2018). Selection signatures will lead to overall improvement in animal productivity and fitness and will contribute ultimately to the economic and food security (Rege and Gibson, 2003). Selection signatures can be used to identify genes and mechanisms associated with different traits such as, *SOCS2* (Yang *et al.*, 2016) and *FGF7* (Gorkhali *et al.*, 2016) genes were found adapted to high-altitude in Tibetan sheep. Genetic structure of sheep tail into long-thin, short-thin, long-fat, short-fat and fat-rumped tail was hypothesized based on the four significant genomic signatures in *HOXB13, TBXT*, *PDGFD* and *IBH* region (Kalds *et al.*, 2022). Moreover, selective sweep studies do not necessarily require phenotypes. Demographic models such as coalescent-based approaches identify sweeps, even in the absence of any observable phenotype in domestic sheep populations (Zhao *et al.*, 2015; Zhang *et al.*, 2021). Skewness of SFS in selected alleles and LD patterns in different populations, gives an understanding of how species evolve and adapt over time (Nielsen, 2001).

Downstream Application

A majority of performance traits are complex and multigenic (Yang *et al.* 2011, Andersson and Georges, 2004), and difficult to modulate directly with biotechnological tools such as genome editing (Blighe *et al.* 2018). Thus, selective sweeps help in fine-tuning the number of genomic regions routed for further validation of causal association with traits of interest using GWAS, qPCR and gene editing. The strength of selection studies is that they are powerful for identifying candidates for further testing. Provided a genotype–phenotype map for non-model organisms is challenging if sample sizes necessary for GWAS are unattainable, biotechniques can serve as an alternative (Rocha *et al.*, 2021) to study phenotypes expressed at the cellular level using cell culture, and *CRISPR-Cas9* (Knott and Doudna, 2018). *LEPR*, *LDB1*, *EGFR* and *FGF2* genes have found to be selectively swept followed by qPCR validation for high-altitude adaptation in Tibetan and Huanghuai goats (Jin *et al.*, 2020). A strong selective signature at the causal mutation (c.334C > A, p.G112W) in *TBXT* gene has confirmed its association with tail length in an inter-cross population of 110 F_2 offspring with varied number of vertebrae by whole-genome association analysis and CRISPR-Cas9 genome editing (Li *et al.*, 2022).Though CRISPR/Cas9 has been used to achieve muscle hypertrophy in sheep (Wang *et al.*, 2016), increase wool and hair length in sheep and goats (Wang *et al.*, 2016), lysozyme coding (Brundige *et al.*, 2008) and biopharming of spider silk in goat milk (Majumder *et al.*, 2015), the potential of CRISPR/Cas9 to edit genes swept under selection is untapped in small ruminants.

The Way Forward

The natural and artificial selection of sheep and goats under different agro-climates has led to unique genes and haplotypes with desirable phenotypes for production, reproduction, disease resistance/resilience and fitness. Selective sweeps identified within a breed and between breeds can be used for selective breeding and crossbreeding, respectively. Different algorithms and software have surfaced to survey the selective sweeps. Often, all the assumptions of statistical models are not met which calls for inclusion of machine learning for identifying selection signatures (Kumar *et al.*, 2022). Fine-tuning the sweep window by pin-pointing selective sweeps (Delaneau *et al.*, 2007) can be done. However, there are still variants other than SNPs under selection such as copy number variations (CNVs) (Zhang *et al.*, 2020), insertions-deletions (InDels) (Boschiero *et al.*, 2018), microsatellites (SSRs) (Clop *et al.*, 2006), minisatellites (Kokita *et al.*, 2013) and epigenetic variants (Aref-Eshghi *et al.*, 2018). Moreover, focus should be given to other forms of selection including negative directional, balancing and disruptive selection in addition to positive selection (Sarvanan *et al.*, 2020). This will help explore genetic diction to make informed breeding decisions for enhanced productivity of small ruminants in their native environment vis-a-vis food security and climate change.

References

Andersson, L., & Georges, M. (2004). Domestic-animal genomics: deciphering the genetics of complex traits. Nature Reviews Genetics, 5(3), 202-212.

Aref-Eshghi, E., Rodenhiser, D. I., Schenkel, L. C., Lin, H., Skinner, C., Ainsworth, P., ... &Sadikovic, B. (2018). Genomic DNA methylation signatures enable concurrent diagnosis and clinical genetic variant classification in neurodevelopmental syndromes. The American Journal of Human Genetics, 102(1), 156-174.

Avise, J. C., & Ayala, F. J. (2009). In the light of evolution III: Two centuries of Darwin. Proceedings of the National Academy of Sciences, 106(supplement_1), 9933–9938.

Alberto, F. J., Boyer, F., Orozco-terWengel, P., Streeter, I., Servin, B., de Villemereuil, P., Benjelloun, B., Librado, P., Biscarini, F., Colli, L., Barbato, M., Zamani, W., Alberti, A., Engelen, S., Stella, A., Joost, S., Ajmone-Marsan, P., Negrini, R., Orlando, L., … Pompanon, F. (2018). Convergent genomic signatures of domestication in sheep and goats. Nature Communications, 9(1), Article 1.https://doi.org/10.1038/s41467-018-03206-y

Beaumont, M. A. (2005). Adaptation and speciation: What can Fst tell us? Trends in Ecology & Evolution, 20(8), 435–440.

Berry, A. J., Ajioka, J. W., &Kreitman, M. (1991). Lack of polymorphism on the Drosophila fourth chromosome resulting from selection. Genetics, 129(4), 1111-1117.

Blighe, K., DeDionisio, L., Christie, K. A., Chawes, B., Shareef, S., Kakouli-Duarte, T., ... & Moore, C. B. T. (2018). Gene editing in the context of an increasingly complex genome. BMC genomics, 19(1), 1-20.

Blott, S. C., Williams, J. L., & Haley, C. S. (1998). Genetic relationships among European cattle breeds. Animal Genetics, 29(4), 273–282.

Boschiero, C., Moreira, G. C. M., Gheyas, A. A., Godoy, T. F., Gasparin, G., Mariani, P. D. S. C., & Coutinho, L. L. (2018). Genome-wide characterization of genetic variants and putative regions under selection in meat and egg-type chicken lines. BMC genomics, 19, 1-18.

Braverman, J. M., Hudson, R. R., Kaplan, N. L., Langley, C. H., & Stephan, W. (1995). The hitchhiking effect on the site frequency spectrum of DNA polymorphisms. Genetics, 140(2), 783–796.

Brundige, D. R., Maga, E. A., Klasing, K. C., & Murray, J. D. (2008). Lysozyme transgenic goats' milk influences gastrointestinal morphology in young pigs. The Journal of Nutrition, 138(5), 921–926.

Charlesworth, B., Morgan, M. T., & Charlesworth, D. (1993). The effect of deleterious mutations on neutral molecular variation. Genetics, 134(4), 1289–1303.

Charlesworth, D. (2006). Balancing selection and its effects on sequences in nearby genome regions. PLoS Genetics, 2(4), e64.

Chen, Q., Chai, Y., Zhang, W., Cheng, Y., Zhang, Z., An, Q., Chen, S., Man, C., Du, L., Zhang, W., & Wang, F. (2022a). Whole-Genome Sequencing Reveals the Genomic Characteristics and Selection Signatures of Hainan Black Goat. Genes, 13(9), Article 9.https://doi.org/10.3390/genes13091539

Chen, Y., Li, R., Sun, J., Li, C., Xiao, H., & Chen, S. (2022b). Genome-Wide Population Structure and Selection Signatures of Yunling Goat Based on RAD-seq. Animals, 12(18), Article 18.https://doi.org/10.3390/ani12182401

Clop, A., Marcq, F., Takeda, H., Pirottin, D., Tordoir, X., Bibé, B., ... & Georges, M. (2006). A mutation creating a potential illegitimate microRNA target site in the myostatin gene affects muscularity in sheep. Nature genetics, 38(7), 813-818.

DAHD (2019). Provisional key results of 20th livestock census. Department of Animal Husbandry and Dairying, Ministry of Fisheries, Animal Husbandry & Dairying, Government of India, Retrieved from http://dahd.nic.in/division/provisionalkey-results-20th-livestock-census

Devendra, C., & Chantalakhana, C. (2002). Animals, poor people and food insecurity: Opportunitics for improved livelihoods through efficient natural resource management. Outlook on Agriculture, 31(3), 161–175.

Falconer, D. S. (1996). Introduction to quantitative genetics. Pearson Education India.

Fan, S., Hansen, M. E., Lo, Y., &Tishkoff, S. A. (2016). Going global by adapting local: A review of recent human adaptation. Science, 354(6308), 54–59.

FAOSTAT. (2021). Retrieved February 9, 2023, from https://www.fao.org/faostat/en/#data/QCL

Fay, J. C., & Wu, C.-I. (2000). Hitchhiking under positive Darwinian selection. Genetics, 155(3), 1405–1413.

Ferrer-Admetlla, A., Liang, M., Korneliussen, T., & Nielsen, R. (2014). On detecting incomplete soft or hard selective sweeps using haplotype structure. Molecular Biology and Evolution, 31(5), 1275–1291.

Ferrer-Admetlla, A., Liang, M., Korneliussen, T., & Nielsen, R. (2014). On Detecting Incomplete Soft or Hard Selective Sweeps Using Haplotype Structure. Molecular Biology and Evolution, 31(5), 1275–1291.https://doi.org/10.1093/molbev/msu077

Field, Y., Boyle, E. A., Telis, N., Gao, Z., Gaulton, K. J., Golan, D., Yengo, L., Rocheleau, G., Froguel, P., & McCarthy, M. I. (2016). Detection of human adaptation during the past 2000 years. Science, 354(6313), 760–764.

Foll, M., &Gaggiotti, O. (2008). A genome-scan method to identify selected loci appropriate for both dominant and codominant markers: A Bayesian perspective. Genetics, 180(2), 977–993.

Gorkhali, N. A., Dong, K., Yang, M., Song, S., Kader, A., Shrestha, B. S., ... & Ma, Y. (2016). Genomic analysis identified a potential novel molecular mechanism for high-altitude adaptation in sheep at the Himalayas. Scientific Reports, 6(1), 29963.

Gouveia, J. J. de S., Silva, M. V. G. B. da, Paiva, S. R., & Oliveira, S. M. P. de. (2014). Identification of selection signatures in livestock species. Genetics and Molecular Biology, 37, 330–342.

Grossman, S. R., Shylakhter, I., Karlsson, E. K., Byrne, E. H., Morales, S., Frieden, G., Hostetter, E., Angelino, E., Garber, M., &Zuk, O. (2010). A composite of multiple signals distinguishes causal variants in regions of positive selection. Science, 327(5967), 883–886.

Guan, D., Luo, N., Tan, X., Zhao, Z., Huang, Y., Na, R., Zhang, J., & Zhao, Y. (2016). Scanning of selection signature provides a glimpse into important economic traits in goats (Capra hircus). Scientific Reports, 6(1), 36372.https://doi.org/10.1038/srep36372

Guang-Xin, Zhao, Y. J., & Huang, Y.-F. (2019). Selection signatures of litter size in Dazu black goats based on a whole genome sequencing mixed pools strategy. Molecular Biology Reports, 46(5), 5517–5523.https://doi.org/10.1007/s11033-019-04904-6

Haldane, J. B. S. (1926). A mathematical theory of natural and artificial selection. Mathematical Proceedings of the Cambridge Philosophical Society, 23(4), 363–372.

Harris, E. E., & Meyer, D. (2006). The molecular signature of selection underlying human adaptations. American Journal of Physical Anthropology, 131(S43), 89-130.

Hermisson, J., &Pennings, P. S. (2005). Soft sweeps: Molecular population genetics of adaptation from standing genetic variation. Genetics, 169(4), 2335–2352.

Hermisson, J., &Pennings, P. S. (2017). Soft sweeps and beyond: Understanding the patterns and probabilities of selection footprints under rapid adaptation. Methods in Ecology and Evolution, 8(6), 700–716.

Hohenlohe, P. A., Phillips, P. C., &Cresko, W. A. (2010). Using population genomics to detect selection in natural populations: Key concepts and methodological considerations. International Journal of Plant Sciences, 171(9), 1059–1071.

Hussin, J., Nadeau, P., Lefebvre, J.-F., &Labuda, D. (2010). Haplotype allelic classes for detecting ongoing positive selection. BMC Bioinformatics, 11, 1–11.

Islam, R., Li, Y., Liu, X., Berihulay, H., Abied, A., Gebreselassie, G., Ma, Q., & Ma, Y. (2019). Genome-Wide Runs of Homozygosity, Effective Population Size, and Detection of Positive Selection Signatures in Six Chinese Goat Breeds. Genes, 10(11), Article 11.https://doi.org/10.3390/genes10110938

Jin, M., Lu, J., Fei, X., Lu, Z., Quan, K., Liu, Y., Chu, M., Di, R., Wang, H., & Wei, C. (2020). Genetic Signatures of Selection for Cashmere Traits in Chinese Goats. Animals, 10(10), Article 10.https://doi.org/10.3390/ani10101905

Kalds, P., Huang, S., Chen, Y., & Wang, X. (2022). Ovine HOXB13: expanding the gene repertoire of sheep tail patterning and implications in genetic improvement. Communications Biology, 5(1), 1196.

Kaplan, N. L., Hudson, R. R., & Langley, C. H. (1989). The" hitchhiking effect" revisited. Genetics, 123(4), 887–899.

Kim, E.-S., Elbeltagy, A. R., Aboul-Naga, A. M., Rischkowsky, B., Sayre, B., Mwacharo, J. M., & Rothschild, M. F. (2016). Multiple genomic signatures of selection in goats and sheep indigenous to a hot arid environment. Heredity, 116(3), Article 3.https://doi.org/10.1038/hdy.2015.94

Kim, Y., & Nielsen, R. (2004). Linkage disequilibrium as a signature of selective sweeps. Genetics, 167(3), 1513–1524.

Kim, Y., & Stephan, W. (2002). Detecting a local signature of genetic hitchhiking along a recombining chromosome. Genetics, 160(2), 765–777.

Knott, Gavin J., and Jennifer A. Doudna. "CRISPR-Cas guides the future of genetic engineering." Science 361.6405 (2018): 866-869.

Kokita, T., Takahashi, S., &Kumada, H. (2013). Molecular signatures of lineage-specific adaptive evolution in a unique sea basin: the example of an anadromous goby L eucopsarionpetersii. Molecular ecology, 22(5), 1341-1355.

Kosgey, I. S., &Okeyo, A. M. (2007). Genetic improvement of small ruminants in low-input, smallholder production systems: Technical and infrastructural issues. Small Ruminant Research, 70(1), 76–88.

Kreitman, M. (2000). Methods to detect selection in populations with applications to the human. Annual Review of Genomics and Human Genetics, 1(1), 539–559.

Kumar, H., Panigrahi, M., Panwar, A., Rajawat, D., Nayak, S. S., Saravanan, K. A., ... &Dutt, T. (2022). Machine-learning prospects for detecting selection signatures using population genomics data. Journal of Computational Biology, 29(9), 943-960.Delaneau, O., Coulonges, C., Boelle, P. Y., Nelson, G., Spadoni, J. L., &Zagury, J. F. (2007). ISHAPE: new rapid and accurate software for haplotyping. BMC bioinformatics, 8(1), 1-13.

Lewontin, R. C., & Krakauer, J. (1973). Distribution of gene frequency as a test of the theory of the selective neutrality of polymorphisms. Genetics, 74(1), 175–195.

Li, X., He, S. G., Li, W. R., Luo, L. Y., Yan, Z., Mo, D. X., ... & Li, M. H. (2022). Genomic analyses of wild argali, domestic sheep, and their hybrids provide insights into chromosome evolution, phenotypic variation, and germplasm innovation. Genome Research, 32(9), 1669-1684.

Ma, Y., Ding, X., Qanbari, S., Weigend, S., Zhang, Q., &Simianer, H. (2015). Properties of different selection signature statistics and a new strategy for combining them. Heredity, 115(5), 426–436.

Majumder, S., Kaulaskar, M. D., &Neogi, S. (2015). "Biosteel": An exciting product from nature that is superior to many manmade alternatives. Reviews in Chemical Engineering, 31(5), 509–519.

McVean, G. (2007). The structure of linkage disequilibrium around a selective sweep. Genetics, 175(3), 1395-1406.

Mirkena, T., Duguma, G., Haile, A., Tibbo, M., Okeyo, A. M., Wurzinger, M., &Sölkner, J. (2010). Genetics of adaptation in domestic farm animals: A review. Livestock Science, 132(1–3), 1–12.

Mohamadipoor Saadatabadi, L., Mohammadabadi, M., Amiri Ghanatsaman, Z., Babenko, O., Stavetska, R., Kalashnik, O., Kucher, D., Kochuk-Yashchenko, O., &AsadollahpourNanaei, H. (2021). Signature selection analysis reveals candidate genes associated with production traits in Iranian sheep breeds. BMC Veterinary Research, 17(1), 369.https://doi.org/10.1186/s12917-021-03077-4

Morgan, T. H. (1911). Random segregation versus coupling in Mendelian inheritance. Science, 34(873), 384–384.

Nei, M. (2005). Selectionism and neutralism in molecular evolution. Molecular biology and evolution, 22(12), 2318-2342.

Nicolazzi, E. L., Biffani, S., Biscarini, F., Orozco terWengel, P., Caprera, A., Nazzicari, N., & Stella, A. (2015). Software solutions for the livestock genomics SNP array revolution. Animal Genetics, 46(4), 343–353.

Nie, Y., Li, S., Zheng, X., Chen, W., Li, X., Liu, Z., Hu, Y., Qiao, H., Qi, Q., & Pei, Q. (2018). Transcriptome reveals long non-coding RNAs and mRNAs involved in primary wool follicle induction in carpet sheep fetal skin. Frontiers in Physiology, 9, 446.

Nielsen, R. (2001). Statistical tests of selective neutrality in the age of genomics. Heredity, 86(6), 641-647.

Oleksyk, T. K., Smith, M. W., & O'Brien, S. J. (2010). Genome-wide scans for footprints of natural selection. Philosophical Transactions of the Royal Society B: Biological Sciences, 365(1537), 185–205.

Onzima, R. B., Upadhyay, M. R., Doekes, H. P., Brito, Luiz. F., Bosse, M., Kanis, E., Groenen, M. A. M., &Crooijmans, R. P. M. A. (2018). Genome-Wide Characterization of Selection Signatures and Runs of Homozygosity in Ugandan Goat Breeds. Frontiers in Genetics, 9.https://www.frontiersin.org/articles/10.3389/fgene.2018.00318

Pennings, P. S., &Hermisson, J. (2006). Soft sweeps II—molecular population genetics of adaptation from recurrent mutation or migration. Molecular Biology and Evolution, 23(5), 1076–1084.

Pritchard, J. K., Pickrell, J. K., & Coop, G. (2010). The genetics of human adaptation: hard sweeps, soft sweeps, and polygenic adaptation. Current biology, 20(4), R208-R215.

Przeworski, M. (2002). The signature of positive selection at randomly chosen loci. Genetics, 160(3), 1179-1189.

Randhawa, I. A. S., Khatkar, M. S., Thomson, P. C., &Raadsma, H. W. (2014). Composite selection signals can localize the trait specific genomic regions in multi-breed populations of cattle and sheep. BMC genetics, 15(1), 1-19.

Rege, J. E. O., & Gibson, J. P. (2003). Animal genetic resources and economic development: Issues in relation to economic valuation. Ecological Economics, 45(3), 319–330.

Rocha, J. L., Godinho, R., Brito, J. C., & Nielsen, R. (2021). Life in deserts: the genetic basis of mammalian desert adaptation. Trends in ecology & evolution, 36(7), 637-650.

Sabeti, P. C., Reich, D. E., Higgins, J. M., Levine, H. Z., Richter, D. J., Schaffner, S. F., Gabriel, S. B., Platko, J. V., Patterson, N. J., & McDonald, G. J. (2002). Detecting recent positive selection in the human genome from haplotype structure. Nature, 419(6909), 832–837.

Saravanan, K. A., Panigrahi, M., Kumar, H., Bhushan, B., Dutt, T., & Mishra, B. P. (2020). Selection signatures in livestock genome: A review of concepts, approaches and applications. Livestock Science, 241, 104257.

Saravanan, K. A., Panigrahi, M., Kumar, H., Bhushan, B., Dutt, T., & Mishra, B. P. (2021). Genome-wide analysis of genetic diversity and selection signatures in three Indian sheep breeds. Livestock Science, 243, 104367.https://doi.org/10.1016/j.livsci.2020.104367

Schaad, E. W., & Poe, S. (2010). Patterns of ecomorphological convergence among mainland and island Anolis lizards. Biological Journal of the Linnean Society, 101(4), 852–859.

Schluter, D. (1988). Character displacement and the adaptive divergence of finches on islands and continents. The American Naturalist, 131(6), 799–824.

Schlötterer, C. (2003). Hitchhiking mapping–functional genomics from the population genetics perspective. Trends in Genetics, 19(1), 32-38.

Serranito, B., Cavalazzi, M., Vidal, P., Taurisson-Mouret, D., Ciani, E., Bal, M., Rouvellac, E., Servin, B., Moreno-Romieux, C., Tosser-Klopp, G., Hall, S. J. G., Lenstra, J. A., Pompanon, F., Benjelloun, B., & Da Silva, A. (2021). Local adaptations of Mediterranean sheep and goats through an integrative approach. Scientific Reports, 11(1), Article 1.https://doi.org/10.1038/s41598-021-00682-z

Serranito, B., Taurisson-Mouret, D., Harkat, S., Laoun, A., Ouchene-Khelifi, N.-A., Pompanon, F., Benjelloun, B., Cecchi, G., Thevenon, S., Lenstra, J. A., & Da Silva, A. (2021). Search for Selection Signatures Related to Trypanosomosis Tolerance in African Goats. Frontiers in Genetics, 12.https://www.frontiersin.org/articles/10.3389/fgene.2021.715732

Sharma, H., Sharma, R., Ahlawat, S., Das, P., Jayakumar, S., &Tantia, M. (2018). Cattle microsatellite markers successfully established diversity status of Arunachali yak (only registered yak breed of India). Indian Journal of Animal Sciences, 88(9), 1051–1057.

Smith, J. M., & Haigh, J. (1974). The hitch-hiking effect of a favourable gene. Genetics Research, 23(1), 23–35.

Storz, J. F. (2005). invited review: Using genome scans of DNA polymorphism to infer adaptive population divergence. Molecular Ecology, 14(3), 671–688.

Tajima, F. (1989). Statistical method for testing the neutral mutation hypothesis by DNA polymorphism. Genetics, 123(3), 585–595.

Tang, K., Thornton, K. R., &Stoneking, M. (2007). A new approach for using genome scans to detect recent positive selection in the human genome. PLoS Biology, 5(7), e171.

Tao, L., He, X., Jiang, Y., Liu, Y., Ouyang, Y., Shen, Y., Hong, Q., & Chu, M. (2021). Genome-Wide Analyses Reveal Genetic Convergence of Prolificacy between Goats and Sheep. Genes, 12(4), Article 4.https://doi.org/10.3390/genes12040480

Templeton, A. R. (2009). Natural selection from Darwin to the 21st century. Israel Journal of Ecology & Evolution, 55(3), 207–214.

Utsunomiya, Y. T., Pérez O'Brien, A. M., Sonstegard, T. S., Van Tassell, C. P., do Carmo, A. S., Meszaros, G., ... & Garcia, J. F. (2013). Detecting loci under recent positive selection in dairy and beef cattle by combining different genome-wide scan methods. PloS one, 8(5), e64280.

Vanzampuii, S. T. (2022). Whole genome sequencing of Changthangi goat for identification of selection footprints [Unpublished master's thesis]. National Dairy Research Institute, India.

Voight, B. F., Kudaravalli, S., Wen, X., & Pritchard, J. K. (2006). A map of recent positive selection in the human genome. PLoS Biology, 4(3), e72.

Wang, X., Liu, J., Zhou, G., Guo, J., Yan, H., Niu, Y., Li, Y., Yuan, C., Geng, R., Lan, X., An, X., Tian, X., Zhou, H., Song, J., Jiang, Y., & Chen, Y. (2016). Whole-genome sequencing of eight goat populations for the detection of selection signatures underlying production and adaptive traits. Scientific Reports, 6(1), 38932.https://doi.org/10.1038/srep38932

Wang, X., Niu, Y., Zhou, J., Yu, H., Kou, Q., Lei, A., Zhao, X., Yan, H., Cai, B., & Shen, Q. (2016). Multiplex gene editing via CRISPR/Cas9 exhibits desirable muscle hypertrophy without detectable off-target effects in sheep. Scientific Reports, 6(1), 32271.

Weigand, H., & Leese, F. (2018). Detecting signatures of positive selection in non-model species using genomic data. Zoological Journal of the Linnean Society, 184(2), 528–583.

Weir, B. S. (2008). Linkage disequilibrium and association mapping. Annu. Rev. Genomics Hum. Genet., 9, 129–142.

Whitlock, M. C. (2005). Combining probability from independent tests: the weighted Z-method is superior to Fisher's approach. Journal of evolutionary biology, 18(5), 1368-1373.

World Bank (2017). World development report 2018: Learning to realize education's promise. The World Bank.

Xiong, M., Fan, R., &Jin, L. (2002). Linkage disequilibrium mapping of quantitative trait loci under truncation selection. Human Heredity, 53(3), 158–172.

Yadav, R. (2022). Genomic signatures of meat type goat breeds [Unpublished master's thesis]. National Dairy Research Institute, India.

Yang, J., Manolio, T. A., Pasquale, L. R., Boerwinkle, E., Caporaso, N., Cunningham, J. M., ... & Visscher, P. M. (2011). Genome partitioning of genetic variation for complex traits using common SNPs. Nature genetics, 43(6), 519-525.

Yang, J. I., Li, W. R., Lv, F. H., He, S. G., Tian, S. L., Peng, W. F., ... & Liu, M. J. (2016). Whole-genome sequencing of native sheep provides insights into rapid adaptations to extreme environments. Molecular biology and evolution, 33(10), 2576-2592.

Yaro, M., Munyard, K. A., Morgan, E., Allcock, R. J. N., Stear, M. J., &Groth, D. M. (2019). Analysis of pooled genome sequences from Djallonke and Sahelian sheep of Ghana reveals co-localisation of regions of reduced heterozygosity with candidate genes for disease resistance and adaptation to a tropical environment. BMC Genomics, 20(1), 816. https://doi.org/10.1186/s12864-019-6198-8

Zhang, Y., Hu, Y., Wang, X., Jiang, Q., Zhao, H., Wang, J., ... & Huang, J. (2020). Population structure, and selection signatures underlying high-altitude adaptation inferred from genome-wide copy number variations in Chinese indigenous cattle. Frontiers in genetics, 10, 1404.

Zhang, Y., Xue, X., Liu, Y., Abied, A., Ding, Y., Zhao, S., Wang, W., Ma, L., Guo, J., Guan, W., Pu, Y., Mwacharo, J. M., Han, J., Ma, Y., & Zhao, Q. (2021). Genome-wide comparative analyses reveal selection signatures underlying adaptation and production in Tibetan and Poll Dorset sheep. Scientific Reports, 11(1), Article 1.https://doi.org/10.1038/s41598-021-81932-y

Zhao, F., McParland, S., Kearney, F., Du, L., & Berry, D. P. (2015). Detection of selection signatures in dairy and beef cattle using high-density genomic information. Genetics Selection Evolution, 47(1), 1-12.

9

Artificial Intelligence in Animal Genetics and Breeding

Vikas Vohra, Rani Alex and G.R. Gowane

Animal Genetics and Breeding Division, ICAR-National Dairy Research Institute Karnal, Haryana, India

Artificial intelligence (AI) uses multiple technologies that equip machines to sense, comprehend, plan, act, and learn with human-like levels of intelligence. Fundamentally, AI systems perceive environments, recognize objects, contribute to decision making, solve complex problems, learn from past experiences, and imitate patterns. AI technologies such as deep learning allow computer systems to understand a problem, learn from examples, and make predictions.

The AI has a tremendous potential to solve the diverse application problems in the field of domestic animal breeding and management. Till date, most of the studies have been focused on animal behaviour (Valletta, et al., 2017), cognition (Lake et al., 2017), animal welfare (Neethirajan, 2021), individual identification (Norouzzadeh et al., 2018), growth or body condition scoring (Zin et al., 2020), disease detection and its monitoring (Kumar et al., 2021), precision farm management (Radun et al., 2021) etc. particularly in the pigs, poultry and cattle species. These AI based applications primarily involves data generation, data processing, use of gadgets / IoT devices, development of smart machine learning algorithms and their integration to attain precision to solve various issues in animal husbandry. Having said that, very few works have been carried out with respect to applications of AI in domestic animal breeding, and major problems like breed identification, genomics- based animal selection criteria and breeding decision, are yet to be automated. Moreover, there are some bottlenecks in application of AI in animal breeding, like the lower accuracy of a decision/outcome and large cost involved in creating AI based infrastructure.

Breeding Value Estimation

Breeding Value (BV) of an animal is the foremost genetic parameter used to make the selection decision / continuance of the dairy animal in the herd, i.e. STAY decision of an animal is based on its BV. Assessing the true breeding value for a production and reproduction trait is not feasible, as the trait is governed by many genes, genic interactions, environment and epigenetics. However, an estimate of BV is usually made for a herd or a population. Since, selection of a bull for superior milk production is a sex-limited trait and is not expressed in an individual therefore an estimate of its breeding value (EBV) for milk production is made based on characters measured in sire family. Often these estimates, EBV, is based on multiple sources of information namely, pedigree, phenotype, genotype, physical, body type score etc. EBV of sex-limited traits based on phenotype information has a limited selection efficiency. In other words, the accuracy of a selected animal is less reliable when only the phenotype of a trait is the basis of EBV. To improve the accuracy for prediction of a BV in a dairy herd, the genetic information about the genotypes of candidate genes for milk production were added, i.e. QTL for performance traits were used. These marker loci had a limitation, they were not whole genome based. Very recently, the genomic markers spread across the whole genome were used to predict the breeding value.

Meuwissen (2001) first time attempted to predict total genetic value using genome-wide dense marker maps. He estimated the genomic breeding value (gEBV) based on the effects of approximately 50,000 markers and reported an improve in efficiency of prediction. His landmark work paved the era of genomic prediction and provided proof of concept that selection decision based on gEBV predicted from genome-wide marker-based information could substantially increase the rate of genetic gain in animals. However, the accuracy of prediction is to a large extent depends upon the prediction models using large input information like 50,000 markers and solutions to the mixed model equations.

Selection of Farm Animals

In dairy cattle breeding program, progeny testing program the bulls are evaluated based on the production performance of their daughters. A large time is taken to evaluate a candidate bull as progeny testing programs depends upon generation interval, which is quite long for the dairy bovines. Various prediction models involving the sex-limited dairy traits have been used to predict the animals BV, but none of them is perfect. Different types of information have been used to model the BV, and it's difficult to weigh these traits appropriately. Phenomics and genomic predictions are based on information from multivariable approach. These methods are computationally expensive, time consuming and requires skilled/trained human resource. These factors limit the selection decision based on few morphological traits and finally delays the genetic improvement in dairy animals. Artificial Intelligence can be employed to develop a computational framework for developing an efficient problem-solving method that can make logical deduction and simulate human reasoning to solve the problem of complex trait analysis. Artificial intelligence can assist in selection decisions at dairy farms regarding cattle and buffalo bull at an early age through prediction of accurate estimated breeding value (gEBV). Models used to estimate gEBV in small sized dairy herds are not available at present. AI based methods can help in developing decision support system for a candidate dairy bulls based on extracting relevant genome-based markers.

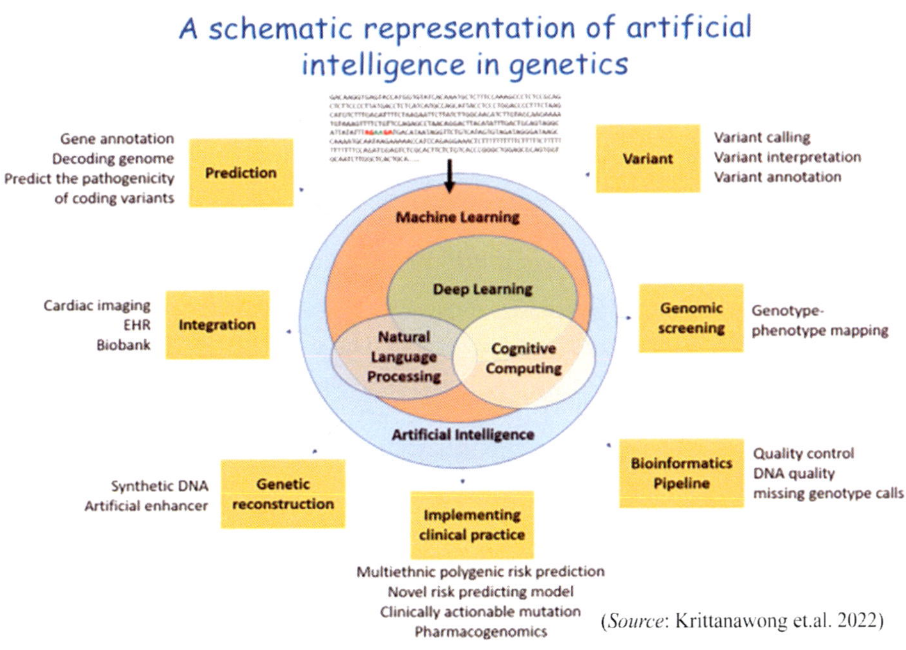

(*Source*: Krittanawong et.al. 2022)

Artificial Intelligence Based Methods in Animal Genetics and Breeding

Superior genetics leads to superior production. The selection decisions in dairy bulls are made on different sources of information like phenotype, additive genetics, environment, genomics, epigenomics etc. In the post genomics era churning big data and its analytics is must. With this aim transforming large amounts of data into valuable knowledge for predicting complex traits has been an important challenge for animal breeders. Artificial intelligence-based learning methods is increasingly used to deal with such problems.

AI efficiently deal with problems in analyzing big data and in situations where the number of parameters is much larger than the number of observations. AI based learning methods learning algorithms such as ensemble methods (random forests (RF) and boosting) and deep learning (DL) algorithms help in genome-enabled prediction. Further, the time taken in decision making (selection or rejection of a bull) for dairy bovines is considerably long and it hinder the pace of genetic improvement. Artificial Intelligence (AI) can be employed to develop a computational framework for developing an efficient problem-solving method that can make logical deduction and simulate human reasoning to solve the problem of complex trait analysis. Artificial intelligence can assist in selection decisions at dairy farms regarding cattle and buffalo bull at an early age through prediction of accurate estimated breeding value (gEBV). However, the models used to estimate gEBV in small sized dairy herds are not available at present. AI learning methods focus on prediction without using a pre-conceived model. On the other hand, conventional statistical approaches formalize relations between variables in the form of explicit mathematical models with parameters that are interpretable in the context of some theory. AI based methods can help in developing decision support system for a candidate dairy bulls based on extracting relevant genome-based markers.

Artificial Intelligence Based Genomic Prediction

The introduction of genomic wide selection by Meuwissen et al. 2001 has revolutionised the animal and plant breeding. The assembly of large reference population is very much required to derive prediction equation (Goddard, 2009; Van Raden et al., 2009), which is very much challenging for small breeds. The predictive ability of the methods also plays a significant role in the accuracy and bias of the predicted breeding values (Tong and Nikoloski, 2021). Along with BLUP, Meuwissen et al. (2001) also introduced other statistical methodological like BayesA, which is a mixed model that assumes each marker is a random effect, and BayesB which a general extension of BayesA by setting a large proportion of SNP markers to be related with null effects. Bayesian approaches have been useful to detect many associations, but significant associations explain only a small fraction of the genetic variance of quantitative traits (Manolio et al., 2009; Visscher et al., 2010). Other methodologies like Naive and Scaled Elastic Net, Least Absolute Angle and Selection Operator (LASSO), Support Vector Regression (SVR) with a linear kernel, Bayesian LASSO (BL), BayesCπ, and BayesDπ (Azodi et al., 2019) were also used for genomic prediction. These all methodologies have good productive ability well when handling high dimensional data structure, but all are based on a linear approximation and therefore fail to capture non-linearity of markers effects such as gene-gene interactions, epistasis and dominance effects, which are probable play important role to explain some variability of complex traits (Monir and Zhu, 2018; Azodi et al., 2019).

The utilization of machine learning approach may make paradigm changes in genetics and breeding and artificial neural networks may be described as future tools in deciding process with great potential in both animal and plant genetics (Gianola et al. 2011; Ventura et al. 2012 and Nascimento et al. 2013). Artificial Neural Networks (ANNs) provides a powerful learning technique on complex features by predicting future results based on training data (Shaneh and Butler, 2006). It uses semi-parametric and nonparametric methods for the genomic-enable prediction of quantitative traits by accounting for non-additive and nonlinear effects as well as genotype-environment interactions (Gianola et al. 2006 and de Campos et al. 2010). Further, the accurate and less biased methodology can be used for development of decisions support system for genomic prediction in bull ranking.

From the introduction of genomic selection, a number of statistical techniques were created to forecast GEBV, among them the Bayesian models and the genomic best linear unbiased prediction (GBLUP) models (Meuwissen et al., 2001; Van Raden, 2008; Habier et al., 2011) have been used extensively. Most of these models departed on the distribution assumption of SNPs. The polygenic nature of the majority of the traits in livestock, ease in its use and lower computational demand have resulted in the popularity of the GBLUP method, (Hayes et al., 2009; Wang et al., 2014; Zhang et al., 2016; Mehrban et al., 2017). Further, Misztal et al., 2009 introduced an approach based on the GBLUP architecture called single-step genomic best linear unbiased prediction (ssGBLUP), which utilises simultaneously all pedigree, genotypic, and phenotypic information from both genotyped and non-genotyped individuals. This approach combines the relationship matrix for the numerator based on pedigree and the relationship matrix for the genotype-based numerator into a single matrix (H) (Misztal et al., 2009; Legrra et al., 2009). Compared to approaches that exclusively employ genotyped individuals, the use of ssGBLUP improves the accuracy of prediction (Misztal et al., 2013), further it is more accurate in comparison to pedigree BLUP as well (Onogi et al., 2014; Ismael et al., 2017). Among tthe Bayesian Alphabet methods, BVS methods such as Bayes-B, were very quickly adapted for use in GWA to differentiate SNPs or windows of SNPs that are associated with phenotype from those that capture just pedigree relationships (Wolc and Dekkers, 2022).

The first application of ANNs for GS was introduced by Okut et al. (2011) to predict body mass index (BMI) in mice. Since then, more complex ANN architectures have been suggested for GS. The neural network architectures such as recurrent neural networks (RNN), long-short term memory (LSTM), convolutional neural network (CNN) and feed forward multilayer perceptron (MLP) has been proved the potential for effective use in GS (Alkhudaydi et al., 2019; Sandhu et al., 2021). As a results of this, several studies have recently been carried out to analyse the use of ANNs for genomic prediction (Bellot et al., 2018 in humans; Ma et al., 2018 in wheat; Waldmann, 2018 in simulated data set of livestock and in pigs; Azodi et al., 2019; Khaki and Wang, 2019 in crops). In India, little work has been carried out for genomic prediction of breeding value of animal using neural networks. That too using Machine learning approaches for phenotype data where ANN based prediction of first lactation 305day milk yield (Sharma et al., 2006) and life time milk production in Sahiwal cattle and Karan Fries cattle (Gandhi et al. 2009 and Chaturvedi et al, 2013) was predicted.

Gianola et al. (2011) used Bayesian ANN architectures using for predicting phenotypes in two data sets consisting of milk production in Jersey cows and yield of inbred lines of wheat and has suggested that neural networks may be useful for predicting complex traits using high-dimensional genomic information, a situation where the number of unknowns exceeds sample size. The genomic prediction performance of support vector regression, random forest regression, reproducing kernel Hilbert space (RKHS), ridge regression, and Bayesian Lasso was compared in maize and wheat datasets (Ornella et al., 2014) with different trait-environment combinations, and suggested that RKHS and random forest regression were the best. Further, González-Camacho et al. (2018) reported that the support vector machine (SVM) with linear kernel performed the best in comparison with other ML methods and linear models in the genomic prediction of the rust resistance of wheat. Li et al. (2020) examined the efficiency of three machine learning methods, namely Random Forests (RF), GBM and XgBoost, for identification of subset of SNPs and use of subset for genomic relationship matrices (GRMs) for the estimation of genomic breeding values using 38,082 SNP markers and body weight phenotypes from 2,093 Brahman cattle. Lopes et al (2020) compared the predictive performance of artificial neural networks (ANNs) with Bayesian ridge regression, Bayesian Lasso, Bayes A, Bayes B and Bayes Cπ in estimating genomic breeding values for meat tenderness in Nellore cattle. Wang et al. (2022) evaluated the efficiency of four machine learning methods, including support vector regression (SVR), kernel ridge regression (KRR), random forest (RF) and Adaboost.R2 for genomic prediction of reproductive traits in Chinese Yorkshire pigs and demonstrated that ML methods especially AdaboostR2_KRR had better overall prediction performance than conventional genomic selection methods, and could be new options for genomic prediction.

Conclusion

Although the accuracy of the various AI methods used till date for genomic evaluations is not significantly higher than conventional models, but the deep learning methods can be very helpful in developing a computational framework to help in Animal breeding decisions.

References

Azodi, C. B., Bolger, E., Mccarren, A., Roantree, M., and Shiu, S. (2019). Benchmarking parametric and machine learning models for genomic prediction of complex traits. G3 9, 3691–3702. doi: 10.1534/g3.119.400498

Alkhudaydi T., Reynolds D., Zhou J., Iglesia B., Griffiths S. (2019). An exploration of deep-learning based phenotypic analysis to detect spike regions in field conditions for UK bread wheat. Plant Phenom. 2019:7368761.

Bellot, P., de Los Campos, G., and Pérez-Enciso, M. (2018). Can deep learning improve genomic prediction of complex human traits? Genetics 210, 809–819. doi: 10.1534/genetics.118.301298

Brito Lopes F, Magnabosco CU, Passafaro TL, Brunes LC, Costa MFO, Eifert EC, Narciso MG, Rosa GJM, Lobo RB, Baldi F. Improving genomic prediction accuracy for meat tenderness in Nellore cattle using artificial neural networks. J Anim Breed Genet. 2020 Sep;137(5):438-448. doi: 10.1111/jbg.12468. Epub 2020 Feb 5. PMID: 32020678.

Chaturvedi, S., Yadav, R.L., Gupta A.K. and Sharma A.K. (2013). Life time milk amount prediction in dairy cows using artificial neural networks. International Journal of Recent Research and Review. 5: 1-6.

De Campos, G., Gianola, D., Rosa, G.J., Weigel, K.A. and Crossa, J. (2010). Semi-parametric genomic enabled prediction of genetic values using reproducing kernel Hilbert spaces methods. Genetic Research. 92: 295-30

Gandhi, R.S., Raja, T.V., Ruhil, A.P. and Kumar A. (2009). Prediction of lifetime milk production using artificial neural network in Sahiwal cattle. Indian Journal of Animal Sciences. 79: 1038-1040.

Gianola, D.,Fernando, R.L. and Stella, A. (2006). Genomic-assisted prediction of genetic value with semiparametric procedures. Genetics. 173: 1761-1776

Gianola, D.H., Okut, K. and G. Rosa. (2011). Predicting complex quantitative traits with Bayesian neural networks: A case study with Jersey cows and wheat. BMC Genetics. 12: 87-96.

Gianola D, Okut H, Weigel KA, Rosa GJ. Predicting complex quantitative traits with Bayesian neural networks: a case study with Jersey cows and wheat. BMC Genet. 2011 Oct 7;12:87. doi: 10.1186/1471-2156-12-87. PMID: 21981731; PMCID: PMC3474182.

Goddard, M. (2009). Genomic selection: prediction of accuracy and maximisation of long-term response. Genetica, 136(2), 245-257.

González-Camacho JM, Ornella L, Pérez-Rodríguez P, Gianola D, Dreisigacker S, Crossa J. Applications of machine learning methods to genomic selection in breeding wheat for rust resistance. Plant Genome. 2018;11(2):170104.

Habier, D., Fernando, R. L., Kizilkaya, K., & Garrick, D. J. (2011). Extension of the Bayesian alphabet for genomic selection. BMC bioinformatics, 12(1), 1-12.

Hayes, B. J., Bowman, P. J., Chamberlain, A. J., & Goddard, M. E. (2009). Invited review: Genomic selection in dairy cattle: Progress and challenges. Journal of dairy science, 92(2), 433-443

Ismael, A., Løvendahl, P., Fogh, A., Lund, M. S., & Su, G. (2017). Improving genetic evaluation using a multitrait single-step genomic model for ability to resume cycling after calving, measured by activity tags in Holstein cows. Journal of dairy science, 100(10), 8188-8196.

Khaki, S., andWang, L. (2019). Crop yield prediction using deep neural networks. Front. Plant Sci. 10:621. doi: 10.3389/fpls.2019.00621

Kumar S., Singh P., Ranjan M. A review on deep learning-based pneumonia detection systems; Proceedings of the 2021 International Conference on Artificial Intelligence and Smart Systems (ICAIS); Coimbatore, India. 25–27 March 2021; pp. 289–296

Lake, B., Ullman, T., Tenenbaum, J., & Gershman, S. (2017). Building machines that learn and think like people. *Behavioral and Brain Sciences, 40*, E253. doi:10.1017/S0140525X16001837

Legarra, A., Aguilar, I., & Misztal, I. (2009). A relationship matrix including full pedigree and genomic information. Journal of Dairy Science, 92(9), 4656-4663.

Li Bo, Zhang Nanxi, Wang You-Gan, George Andrew W., Reverter Antonio, Li Yutao. 2020. Genomic Prediction of Breeding Values Using a Subset of SNPs Identified by Three Machine Learning Methods Frontiers in Genetics 9. https://www.frontiersin.org/articles/10.3389/fgene.2018.00237

Ma, W., Qiu, Z., Song, J., Li, J., Cheng, Q., Zhai, J., et al. (2018). A deep convolutional neural network approach for predicting phenotypes from genotypes. Planta 248, 1307–1318. doi: 10.1007/s00425-018-2976-9

Manolio TA, Collins FS, Cox NJ, Goldstein DB, Hindorff LA, Hunter DJ, McCarthy MI, Ramos EM, Cardon LR, Chakravarti A, Cho JH, Guttmacher AE, Kong A, Kruglyak L, Mardis E, Rotimi CN, Slatkin M, Valle D, Whittemore AS, Boehnke M, Clark AG, Eichler EE, Gibson G, Haines JL, Mackay TF, McCarroll SA, Visscher PM. Finding the missing heritability of complex diseases. Nature. 2009 Oct 8;461(7265):747-53. doi: 10.1038/nature08494. PMID: 19812666; PMCID: PMC2831613.

Mehrban, H., Lee, D. H., Moradi, M. H., IlCho, C., Naserkheil, M., & Ibáñez-Escriche, N. (2017). Predictive performance of genomic selection methods for carcass traits in Hanwoo beef cattle: impacts of the genetic architecture. Genetics Selection Evolution, 49(1), 1-13.

Meuwissen, T. H., Hayes, B. J., & Goddard, M. (2001). Prediction of total genetic value using genome-wide dense marker maps. genetics, 157(4), 1819-1829.

Misztal, I., Aggrey, S. E., & Muir, W. M. (2013). Experiences with a single-step genome evaluation. Poultry science, 92(9), 2530-2534.

Misztal, I., Legarra, A., & Aguilar, I. (2009). Computing procedures for genetic evaluation including phenotypic, full pedigree, and genomic information. Journal of dairy science, 92(9), 4648-4655.

Monir MM and Zhu J (2018) Dominance and Epistasis Interactions Revealed as Important Variants for Leaf Traits of Maize NAM Population. Front. Plant Sci. 9:627. doi: 10.3389/fpls.2018.00627

Nascimento, M., Peternelli, L.A., Cruz, C.D., Nascimento, A.C.C., Ferreira, R.P., Bhering, L.L. and Salgado, C.C. (2013). Artificial neural networks for adaptability and stability evaluation in alfafa genotypes. Crop Breeding and Applied Biotechnology. 12: 152-156

Norouzzadeh Mohammad Sadegh, Nguyen Anh, Kosmala Margaret, Swanson Alexandra, Palmer Meredith S., Packer Craig, and Clune Jeff. 2018. Automatically identifying, counting, and describing wild animals in camera-trap images with deep learning. *Proceedings of the National Academy of Sciences* 115, 25

Onogi, A., Ogino, A., Komatsu, T., Shoji, N., Simizu, K., Kurogi, K., ... & Iwata, H. (2014). Genomic prediction in Japanese Black cattle: application of a single-step approach to beef cattle. Journal of Animal Science, 92(5), 1931-1938.

Ornella L, Perez P, Tapia E, Gonzalez-Camacho JM, Burgueno J, Zhang X, et al. Genomic-enabled prediction with classification algorithms. Heredity (Edinb). 2014;112(6):616–26.

Radun, Viktor & Dokić, Dragan & Gantner, Vesna. (2021). Implementing artificial intelligence as a part of precision dairy farming for enabling sustainable dairy farming. Ekonomika poljoprivrede. 68. 869-880.

Shaneh, A. and Butler, G. (2006). Bayesian learning for feed-forward neural network with application to proteomic data: the glycosylation sites detection of the epidermal growth factor-like proteins associated with cancer as a case study. Proceedings of Conference of the Canadian Society for Computational Studies of Intelligence: 110-121.

Sharma, A.K., Sharma, R.K. and Kasana, H.S. (2006). Empirical comparisons of feed-forward connectionist and conventional regression models for prediction of first lactation 305-day milk yield in Karan Fries dairy cows. Neural Computing and Applications. 15: 359-365.

Suresh Neethirajan (2021). The Use of Artificial Intelligence in Assessing Affective States in Livestock. Frontiers in Veterinary Science, 8, doi=10.3389/fvets.2021.715261

Tong H, Nikoloski Z. Machine learning approaches for crop improvement: Leveraging phenotypic and genotypic big data. J Plant Physiol 2021;257

Valletta JJ, Colin Torney, Michael Kings, Alex Thornton, Joah Madden (2017). Applications of machine learning in animal behaviour studies, *Animal Behaviour* 124:203-220

VanRaden, P. M. (2008). Efficient methods to compute genomic predictions. Journal of Dairy Science, 91(11), 4414-4423.

VanRaden, P. M., Van Tassell, C. P., Wiggans, G. R., Sonstegard, T. S., Schnabel, R. D., Taylor, J. F., & Schenkel, F. S. (2009). Invited review: Reliability of genomic predictions for North American Holstein bulls. Journal of Dairy Science, 92(1), 16-24.

Ventura, R.V., Silva, M.A., Medeiros, T.H., Dionello, N.L., Madalena, F.E., Fridrich, A.B., Valente, B.D., Santos, G.G., Freitas, L.S., Wenceslau, R.R., Felipe, V.P.S. and Corrêa, G.S.S. (2012). Use of artificial neural networks in breeding values prediction for weight at 205 days in Tabapuã beef cattle. Brazilian Archive of Veterinary Medicine and Animal Science. 64: 411-418

Visscher, P. M., Medland, S. E., Ferreira, M. A., Morley, K. I., Zhu, G., Cornes, B. K., ... & Martin, N. G. (2010). Assumption-free estimation of heritability from genome-wide identity-by-descent sharing between full siblings. PLoS Genet, 2(3), e41.

Waldmann, P. (2018). Approximate Bayesian neural networks in genomic prediction. Genet. Select. Evol. 50:70.

Wang, H., Misztal, I., Aguilar, I., Legarra, A., Fernando, R. L., Vitezica, Z., ... & Muir, W. M. (2014). Genome-wide association mapping including phenotypes from relatives without genotypes in a single-step (ssGWAS) for 6-week body weight in broiler chickens. Frontiers in genetics, 5, 134.

Wang, X., Shi, S., Wang, G., … et al. (2022) Using machine learning to improve the accuracy of genomic prediction of reproduction traits in pigs. J Animal Sci Biotechnol 13, 60.

Wolc, A., Dekkers, J.C.M. (2022) Application of Bayesian genomic prediction methods to genome-wide association analyses. Genet Sel Evol 54, 31.

Zhang, Q., Guldbrandtsen, B., Thomasen, J. R., Lund, M. S., & Sahana, G. (2016). Genome-wide association study for longevity with whole-genome sequencing in 3 cattle breeds. Journal of Dairy Science, 99(9), 7289–7298.

Zin, Thi Thi and Seint, Pann Thinzar and Tin, Pyke and Horii, Yoichiro (2020). Body Condition Score Assessment of Depth Image Using Artificial Neural Network. Association for Computing Machinery, Proceedings of the 12th International Conference on Computer Modeling and Simulation, (5):33–37

10

Avenues and Possibilities of Use of Artificial Intelligence (AI) in Animal Breeding in India

Avneesh Kumar[1], Neha Gupta[2], S.P. Singh[1], Rakesh Goel[1] and Deepak Sharma[1]

[1]Department of Animal Genetics & Breeding, Co.V.Sc. & A.H., Uttar Pradesh Pandit Deen Dayal Upadhyaya Pashu Chikitsa Vigyan Vishwavidyalaya Evam Go-Anusandhan Sansthan, Mathura, Uttar Pradesh
[2]Department of Animal Husbandry, Government of Uttar Pradesh

Artificial Intelligence (AI) is a discipline of computer science that now days affecting evary aspect of day to day life. It has the potential to greatly improve animal breeding programs in India by providing more accurate and efficient selection methods. Some of the avenues and possibilities of using AI in animal breeding in India includes Predictive analytics: AI can be used to develop optimization models that take into account a variety of factors, such as genetics, phenotypic traits, and environmental conditions, to identify the best breeding strategies. AI can be used to analyze large amounts of data on animal genetics, phenotypic traits, and environmental factors to predict the likelihood of an animal producing offspring with desired traits. This can help breeders make more informed decisions about which animals to be selected for breeding. AI algorithms can be trained to accurately identify diseases in animals by analyzing imaging data, such as X-rays and MRI scans, as well as blood and tissue samples which in turn facilitate veterinarians to make more accurate diagnoses, improving treatment outcomes. AI can also improved feed efficiency by analyzing the feed consumption and nutrient utilization data to identify patterns and trends that can be further utilized to improve feed efficiency, reduced feed costs and increased production efficiency. AI algorithms can be used to analyze environmental data, such as temperature, humidity and air quality, to identify factors that may impact animal welfare and thus improving the living conditions for animals, better health and well-being. But despite being vastly employed in various areas, industrial, healthcare sectors it is not being fully utilized in animal husbandry. To gain maximum benefit from this emerging field it need to be studied to develop AI based solutions to improve animal husbandry sector of India, but first of all awareness about what is artificial intelligence, what are its type and how it can be achieved should be known which is being discussed in this chapter in following paragraphs.

Artificial Intelligence (AI)

It is a branch of computer science that deals with the development of algorithms and computer systems that can perform tasks that typically require human intelligence, such as recognizing patterns, understanding natural language, making decisions, and solving problems. The ultimate goal of AI research is to create machines that can perform tasks without explicit human instructions, by learning from data and experiences, just like humans do. There are several different subfields of AI, including machine learning, natural language processing, robotics, and computer vision.

Importance of Artificial Intelligence

Artificial Intelligence (AI) has become increasingly important in recent years due to its ability to automate processes and make tasks more efficient. Following are some of the reasons why AI is considered important:

- Improved Accuracy: AI algorithms can analyze large amounts of data and make predictions with a high degree of accuracy, reducing the chances of human error.
- Increased Efficiency: AI-powered systems can process and analyze data much faster than humans, leading to significant time savings and increased efficiency.
- Cost Savings: By automating tasks and reducing the need for manual labor, AI can help companies save money on labor costs.
- Improved Customer Experience: AI can be used to personalize customer experiences, such as through chatbots and personalized recommendations, improving customer satisfaction and loyalty.
- Better Decision Making: AI can help businesses make informed decisions by providing data-driven insights and recommendations.
- Enhanced Productivity: AI can automate repetitive and time-consuming tasks, freeing up employees to focus on more creative and strategic work.

- Predictive Maintenance: AI can help predict when a machine or system is likely to fail, allowing for proactive maintenance and reducing downtime.

Hence we can say that, AI has the potential to transform the way we live and work, and its importance will only continue to grow as technology advances, and it is important to have awareness about various aspects of AI.

Different Types of Artificial Intelligence

Various types of AI represent the progression of AI from simple reactive machines to more advanced systems with greater capabilities. The development of AI continues to be an active area of research, and it is likely that new types of AI will emerge in the future as the field evolves. AI can be broadly categorized into several types based on its capabilities and application. Here are some of the most common types of AI

1. Reactive Machines: Reactive machines are the simplest form of AI. They are designed to respond to specific situations, but they do not have the ability to form memories or make decisions based on past experiences. Examples of reactive machines include Deep Blue, the chess-playing computer developed by IBM.
2. Limited Memory: Limited memory AI systems have the ability to remember past events, but they are not capable of reasoning or making decisions based on that information. For example, self-driving cars use limited memory AI systems to make decisions based on the data collected from their sensors.
3. Theory of Mind: Theory of mind AI systems are capable of understanding and reasoning about mental states, such as beliefs, intentions, and emotions. These systems are still in their infancy, but they hold great promise for applications in fields such as robotics and natural language processing.
4. Self-Aware: Self-aware AI systems are the most advanced form of AI and are capable of introspection and self-reflection. These systems are still purely theoretical and do not yet exist in practice.
5. Narrow AI: Narrow AI refers to AI systems that are designed to perform a specific task, such as image recognition, speech recognition, or playing a game. These systems are trained on a large amount of data and can perform their tasks with high accuracy.

6. General AI: General AI refers to AI systems that are capable of performing a wide range of tasks and are not restricted to a single task. These systems are still purely theoretical and do not yet exist in practice.

Other AI Classifications

In addition to the broad categories of AI mentioned previously, AI can also be classified in other ways based on different aspects such as architecture, techniques, or application domains. Here are some additional AI classifications:

1. Supervised Learning vs Unsupervised Learning: Supervised learning algorithms are trained using labeled data, where the desired output is provided. The algorithm then learns to map inputs to outputs based on this labeled data. Unsupervised learning algorithms, on the other hand, work with unlabeled data and try to find patterns and structure in the data.
2. Reinforcement Learning: Reinforcement learning algorithms are designed to learn from the environment through trial and error. They learn to perform actions that maximize a reward signal and avoid actions that lead to penalties. Reinforcement learning is commonly used in robotics, game playing, and autonomous systems.
3. Deep Learning: Deep learning refers to a subfield of machine learning that uses artificial neural networks with multiple layers to model complex patterns in data. Deep learning algorithms have been successful in a wide range of applications, including image and speech recognition, natural language processing, and recommendation systems.
4. Rule-Based Systems: Rule-based systems use a set of if-then rules to make decisions. The rules are based on expert knowledge and are manually encoded into the system. Rule-based systems are used in applications such as decision support systems and medical diagnosis.
5. Fuzzy Logic Systems: Fuzzy logic systems use fuzzy logic, a mathematical framework for dealing with uncertainty, to make decisions. They can handle imprecise or ambiguous information and are commonly used in control systems and decision-making applications.
6. Evolutionary Algorithms: Evolutionary algorithms are inspired by the process of natural selection and are used for optimization problems. They work by generating a population of potential solutions and evolving them over time through a process of selection and mutation.

Above are some additional AI classifications that can be used to describe and categorize different AI systems based on different aspects. The field of AI is constantly evolving, and new classifications may emerge as the field continues to grow and evolve.

Machine Learning Vs Deep Learning

Artificial intelligence (AI) is a broad field that encompasses several sub-disciplines, including machine learning (ML) and deep learning (DL). Following are the key difference between the ML & DL

Machine learning: It is a type of AI that focuses on developing algorithms that can learn from data and make predictions or take actions based on that data. Machine learning algorithms are designed to identify patterns and relationships in data, and use this information to make predictions or decisions. The algorithms can be supervised, unsupervised, or semi-supervised, depending on the type of data being used.

Deep learning: It is on the other hand, is a type of machine learning that uses artificial neural networks to model complex patterns in data. Deep learning algorithms are designed to learn from large amounts of data and can be used to perform a wide range of tasks, such as image classification, speech recognition, and natural language processing. Unlike traditional machine learning algorithms, deep learning algorithms are designed to automatically learn features from the data, reducing the need for manual feature engineering.

Neural Network

A neural network is a type of machine learning model that is inspired by the structure and function of the human brain. In a neural network, data is processed through a series of interconnected nodes, called artificial neurons, which are organized into layers. Each artificial neuron receives inputs, performs a simple computation on them, and then produces an output that is passed on to other neurons in the network. The computations performed by each neuron are determined by a set of weights and biases, which are adjusted during the training process to minimize the error between the network's predictions and the actual outputs. Neural networks can be used for a wide range of tasks, including image and speech recognition, natural language processing, and decision making. They are particularly useful for problems where there is a lot of complex and non-linear relationships between the inputs and outputs, and where it is difficult to write explicit rules to solve the problem.

Overall, neural networks are a key component of artificial intelligence, and they have been instrumental in driving recent breakthroughs in AI research.

By processing and learning from large amounts of data, neural networks can learn to make predictions and decisions that are comparable or even superior to those made by humans, and they are now widely used in many applications.

Common Type of Neural Networks being Utilised in Animal Husbandry

There are many types of neural networks, and different types of neural networks can be combined or modified in various ways to solve specific problems in animal husbandry but some of the most common ones used in animal husbandry include:

1. Convolutional Neural Networks (CNNs) - these are commonly used for image classification tasks, such as identifying diseases in cattle based on skin lesion images.
2. Recurrent Neural Networks (RNNs) - these are ideal for analyzing time-series data, such as monitoring the activity patterns of dairy cows to identify early signs of illness or heat.
3. Long Short-Term Memory (LSTM) networks - these are a type of RNN that are well-suited for processing sequential data with long-term dependencies, such as predicting the future weight of a pig based on its growth data over time.
4. Artificial Neural Networks (ANNs) - these are basic feed forward networks that can be used for a wide range of tasks, such as classifying the breed of a sheep based on its physical characteristics.
5. Autoencoders - these are unsupervised deep learning algorithms that can be used for dimensionality reduction and anomaly detection, such as detecting abnormal behavior in chickens based on their activity patterns.

Methods to achieve artificial intelligence- There are several methods that can be used to achieve artificial intelligence (AI), including:

1. Rule-based systems: These systems use a set of rules to make decisions or perform tasks. The rules are written by experts in the field and encode their knowledge and expertise. Rule-based systems can be useful for simple, well-defined problems where the rules are straight forward and unambiguous.
2. Expert systems: These are rule-based systems that also use knowledge-based reasoning to make decisions. Expert systems use a database of facts and a set of inference rules to determine the best course of action.

3. Machine learning: This is a subfield of AI that involves training algorithms to learn from data, without being explicitly programmed. There are several different types of machine learning, including supervised learning, unsupervised learning, semi-supervised learning, and reinforcement learning.
4. Deep learning: This is a subfield of machine learning that involves training deep neural networks, which are composed of many interconnected nodes, called artificial neurons. Deep learning is particularly useful for complex, non-linear problems and has achieved state-of-the-art results in tasks such as image and speech recognition, natural language processing, and others.
5. Evolutionary algorithms: These are optimization algorithms that use principles of natural selection and genetics to evolve a population of candidate solutions towards a solution to a problem. Evolutionary algorithms can be used for a wide range of problems, including optimization, search, and pattern recognition.
6. Swarm intelligence: This is a subfield of AI that studies the collective behavior of decentralized systems, such as ant colonies, bird flocks, and fish schools. Swarm intelligence can be used to solve problems that are difficult or impossible for individuals to solve alone, such as optimization, search, and control.

Above are some of the main methods used to achieve artificial intelligence. Each method has its own strengths and limitations, and the choice of method depends on the specific problem and the available resources.

Big Data v/s Artificial Intelligence

Big data and artificial intelligence (AI) are related but distinct concepts. Big data refers to the vast amount of data that are generated every day by individuals, organizations, and devices. This data can come from a variety of sources, including social media, transactional systems, sensors, and more. Big data is characterized by its volume, velocity, variety, and value. Artificial intelligence, on the other hand, refers to the development of computer systems that can perform tasks that typically require human intelligence, such as recognizing patterns, making decisions, and learning from experience. AI can be applied to big data to extract insights and make predictions based on the data. In other words, big data is a resource that can be used to train and evaluate AI models, while AI is a set of techniques that can be used to analyze big data and extract value from it. AI can be used to process big data in real-time, to identify trends and patterns, and to make predictions about future events. So, while big data and AI are related, big data is a resource that can be used to support AI, and AI is a set of techniques that can be used to analyze big data. They complement each other and are both important for organizations that want to gain a competitive advantage through the use of data. In animal husbandry, big data can be generated from a variety of sources, including animal health records, feed and nutrient consumption data, and environmental data, among others. The use of big data in animal husbandry can provide a wealth of information that can be used to improve production efficiency, animal health, and welfare. For example, big data can be used to analyze patterns in feed and nutrient consumption, to identify trends in animal health, and to predict future outcomes, such as growth and development. However, managing and analyzing big data can also be a challenge, as the data may be unstructured, inconsistent, and difficult to interpret. Nevertheless, advances in AI and machine learning are helping to overcome these challenges, and big data is increasingly being used in animal husbandry to inform decision-making and improve outcomes.

Training an AI Model

The requirements for training an AI model vary depending on the complexity of the model and the amount of data used for training, but generally, a powerful computer with a high-performance GPU, large amounts of memory, and fast storage are required. Additionally, specialized software tools and libraries, such as Tensor Flow, Py Torch, or scikit-learn, can be used to simplify the process of building, training, and evaluating AI models. The basic steps involved in training an AI model are as Following:

1. Data collection: This is the first step in training an AI model. The quality and quantity of the data used for training will greatly impact the performance of the model. It's important to ensure that the data is representative of the problem that the model is being trained to solve, and that it is labeled, if required, for supervised learning algorithms.

2. Data preprocessing: This step involves cleaning and transforming the data to make it suitable for training the model. This can include tasks such as removing missing values, normalizing the data, and encoding categorical variables.

3. Model selection: This step involves selecting the type of AI model that is best suited for the problem at hand. There are several types of AI models to choose from, including decision trees, random forests, support vector machines, neural networks, and others.

4. Model training: This is the core step in training an AI model. The model is trained using an optimization algorithm, such as gradient descent, that adjusts the model's parameters, such as weights and biases, to minimize the error between the model's predictions and the actual outputs. The optimization algorithm uses a loss function to quantify the error, and the model's parameters are updated in the direction that minimizes the loss.
5. Model evaluation: This step involves evaluating the performance of the model of data that was not used during training, called the validation set. This helps to assess the generalization performance of the model and to prevent over fitting, which occurs when the model is too closely fitted to the training data.
6. Model tuning: This step involves fine-tuning the model to improve its performance. This can include adjusting the model's hyper parameters, such as the learning rate, the number of hidden layers, or the number of nodes in a neural network, to find the best configuration for the problem.

Reliability, reproducibility and flexibility of artificial intelligence: in perspective of animal breeding- A successful animal breeding program should be reliable, reproducible, and flexible, so that it can continue to produce offspring with desirable traits in response to changing conditions and new information. Such a system can be developed with the help of artificial intelligence-

- Reliability of AI in animal breeding refers to the consistency and dependability of the AI system in accurately predicting and selecting for desired traits in offspring. AI can be used to analyze large amounts of data to identify traits that are associated with desired traits, such as growth rate or disease resistance, for example. A reliable AI system can help to ensure that the best animals are selected for breeding, increasing the chances of producing offspring with desirable traits.
- Reproducibility of AI in animal breeding refers to the ability of different individuals or organizations to obtain consistent results when using the same AI system and data. This is important for validating and comparing AI results, as well as for collaboration between different groups.
- Flexibility of AI in animal breeding refers to the ability of the AI system to adapt and modify its predictions in response to changing conditions and new information. For example, if new information becomes available about a disease that is affecting the breed, the AI system may need to be updated to include this information in its predictions. This is important for ensuring that the AI system continues to provide accurate and useful results in different situations.

The use of AI in animal breeding can bring significant benefits, but it is important to ensure that AI systems are reliable, reproducible, and flexible, so that they can provide accurate and useful results in different situations, for this there is requirement of training the AI model with large amount of data.

Artificial Intelligence (AI) and Biostatistics

Biostatistics is the application of statistical methods to the study of biological phenomena, and AI is increasingly being used to support and enhance various aspects of biostatistics. For example, AI can be used in biostatistics to develop predictive models for disease diagnosis and prognosis, to identify risk factors for diseases, to analyze large and complex biological data sets, and to support drug discovery and development. Biostatistics and AI also intersect in areas such as personalized medicine, where AI can be used to develop customized treatment plans based on an individual's specific genetic and molecular profile, and in clinical trial design, where AI can be used to optimize the selection of participants and to improve the accuracy of results. In particular, machine learning algorithms, a subfield of AI, can be trained on large amounts of biological data to identify patterns and make predictions that would be difficult or impossible for humans to detect. The integration of AI and biostatistics had greatly advanced our understanding of biological phenomenon and improved health outcomes, improved agriculture and improved animal husbandry.

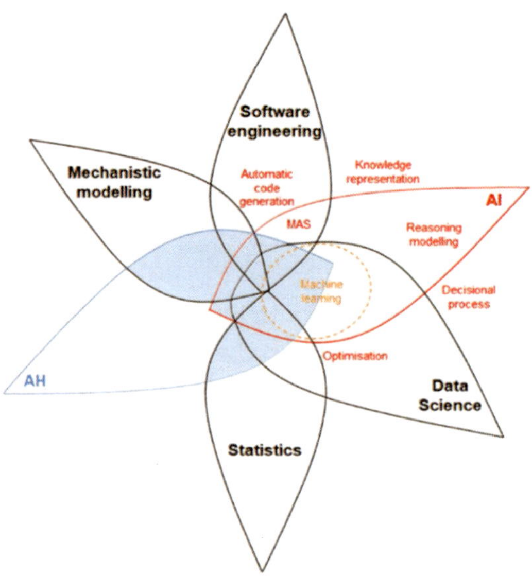

Fig. Interactions between animal health (AH), artificial intelligence (AI), and closely related research domains (Ezanno et al 2021)

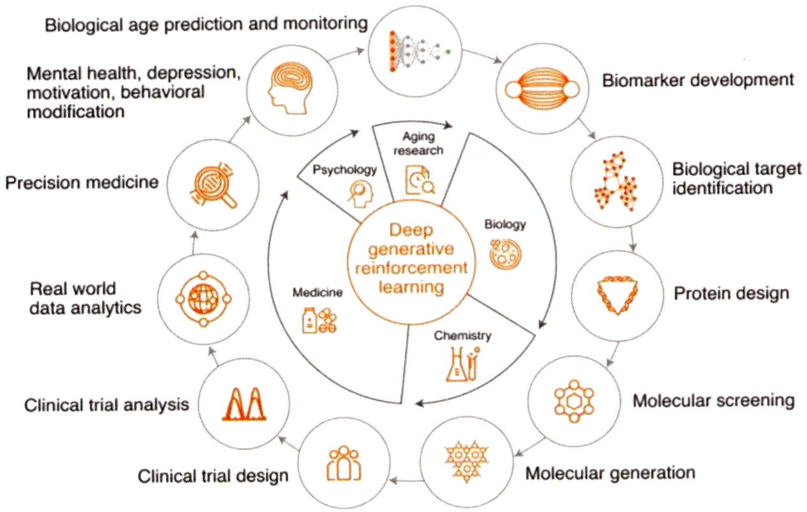

FIG. Applications of ai in health sector which can be extended to veterinary care sector (Zhavoronkov *et al*, 2021)

Integration of AI with Various Omics Studies

Now a day the integration of AI with various omics studies, such as genomics, transcriptomics, proteomics, and metabolomics, is rapidly advancing the field of systems biology. The integration of AI and omics data allows for the analysis of large and complex data sets, providing new insights into the underlying biological processes that contribute to health and disease. For example, in genomics, AI algorithms can be used to analyze large DNA sequencing data sets. In transcriptomics, AI algorithms can be used to analyze large amounts of RNA sequencing data to identify changes in gene expression. In proteomics, AI algorithms can be used to predict protein structures and functions, and to identify potential drug targets. In metabolomics, AI algorithms can be used to identify changes in metabolic pathways that are associated with disease, and to develop new diagnostic and therapeutic strategies. It would not be an exaggeration to say that the integration of AI with omics data has the potential to revolutionize our understanding of biological systems and improve human and animal health.

Genomics and AI (Artificial Intelligence) are two rapidly growing fields that are increasingly intersecting. Genomics is the study of the genetic makeup of living organisms, including the sequencing and analysis of DNA. AI, on the other hand, is the simulation of human intelligence in machines that are capable of performing tasks that typically require human intelligence such as visual perception, speech recognition, decision-making, and language translation.

Use of Artificial Intelligence in Genomics

In recent years, the advancement of DNA sequencing technology has generated a large amount of genetic data, making genomics a rich source of information for AI algorithms. By integrating AI techniques with genomics, scientists are able to analyze vast amounts of genetic data, identify new genomic markers, and develop more accurate disease diagnoses and treatments. Large genomic datasets have been analysed by researchers to identify genetic variants that are associated with certain diseases which led to the development of more personalized and effective treatments for patients. Additionally, AI can be used to predict how a animal will respond to a particular drug based on their genetic information.

It is imperative to note that, the intersection of genomics and AI holds great promise for advancing our understanding of genetics and system biology.

Applications of Artificial Intelligence in the Field of Animal Genetics and Breeding

It is to be familiar with artificial intelligence (AI) and its applications can bring numerous benefits in the field of animal genetics and breeding. Following are some reasons why studying and being aware of AI is important for animal geneticists and breeders:

1. Improved Breeding Decisions: AI can be used to analyze large amounts of genetic data to identify the most desirable traits in animals, such as disease resistance, growth rate, and meat quality. This information can be used to make informed breeding decisions and improve the genetic quality of animal populations.
2. Predictive Breeding: AI algorithms can be used to predict the likelihood of an animal developing certain traits based on its genetic information. This can help breeders make informed decisions about which animals to breed, and avoid breeding animals that are likely to have undesirable traits.
3. Enhanced Trait Selection: AI can be used to identify the specific genetic variations that are associated with desirable traits, such as improved feed efficiency or increased milk production. This information can be used to make more precise selections in breeding programs, leading to faster progress in improving the genetic quality of animal populations.
4. Improved Data Management: AI can be used to manage and analyze large amounts of genetic and phenotypic data, making it easier for

breeders to access and use the information they need. This can improve the efficiency of record keeping and reduce the likelihood of errors.

5. Research Support: AI can be used to support animal genetic research, for example, by analyzing large amounts of data from genomic studies and identifying new genetic variations associated with important traits. This can lead to a deeper understanding of animal genetics and inform the development of new breeding programs.

In conclusion, being familiar with AI and its applications can bring numerous benefits to the fields of animal genetics and breeding. By improving breeding decisions, predictive breeding, enhanced trait selection, improved data management, and research support, AI can help animal geneticists and breeders improve the genetic quality of animal populations and advance the field of animal genetics.

Ethical Consideration of Use of AI in Animal Genetics and Breeding

It's important to consider following issues to develop ethical guidelines and regulations to ensure that the technology is used in a responsible and sustainable manner-

1. Animal welfare: The use of AI in animal breeding may result in animals being subjected to genetic manipulations that can affect their health and welfare. It's important to consider the ethical implications of these practices and to ensure that animals are not subjected to undue harm.
2. Privacy and data protection: AI systems used in animal genetics and breeding rely on large amounts of data, including genetic information about the animals. It's important to consider the intellectual property rights and data protection related to animals and their owners, and to ensure that the data is collected and used in a responsible and ethical manner.
3. Responsibility and accountability: The use of AI in animal genetics and breeding can have unintended consequences, and it's important to consider who is responsible and accountable for these outcomes. For example, if an AI system leads to the production of animals with health problems, who is responsible for addressing these issues?
4. Bias and fairness: AI systems can be biased and unfair, and this can have implications for animal genetics and breeding. For example, if an AI system is biased towards certain traits, it may result in the overproduction of animals with these traits, which could have negative consequences for the animals and farming system.

5. Environmental impact: The use of AI in animal genetics and breeding can also have implications for the environment. For example, if AI systems should be devised to optimize livestock production, in such a way it should not have a negative impact on the environment.

References

Ezanno, P., Picault, S., Beaunée, G. et al. Research perspectives on animal health in the era of artificial intelligence. Vet Res 52, 40 (2021). https://doi.org/10.1186/s13567-021-00902-4

https://en.wikipedia.org/wiki/Artificial_intelligence

Zhavoronkov, A., Bischof, E. & Lee, KF. Artificial intelligence in longevity medicine. Nat Aging 1, 5–7 (2021). https://doi.org/10.1038/s43587-020-00020-4

Part B
Improving the Performance of Livestock Using Genetic Tools

11

Sex Pre-Selection in Farm Animals From Sex Sorting to Sex Ratio Enrichment

Ameya Santhosh[1] and Nihar Ranjan Sahoo[2]

[1]Department of Animal Genetics and Breeding, National Dairy Research Institute Karnal, Haryana
[2]International Centre for FMD, ICAR-Directorate of Foot and Mouth Disease Arugul, Bhubaneswar, Odisha

Introduction

The sex of the animal is one of the most easily observable attributes of a livestock population. The importance of the sex ratio lies in the fact that it sets the future rates of fertility, mortality, and migration. In livestock, male animals are required in fewer numbers compared to females for breeding. So, for profitable management of a farm, the sex ratio has been fixed for each species. But because of the natural 50:50 chance of both types of sperms to fertilize an ovum, this preferred sex ratio is never achieved. In order to reach this economical ratio, additional efforts must be done in the form of culling. Moreover, in monotocous species with nearly one year gestation period, a long wait is sometimes required for getting the offspring of particular sex.

In dairy farming, the number of female offspring born is always an asset to the farm. For the species reared for meat, the males are often avoided due to the intense odor of their meat and handling difficulty while rearing. But when there is shortage of good quality breeding bulls, the sorted sperms will be helpful for producing the elite bulls for breeding. If we can produce only the animal of the desired gender, we can eliminate split-sex feeding, reduce problems related to animal welfare and castration, improve production efficiencies. Additionally, the costs of the progeny testing programme and embryo transfer are reduced as a result, and the value of genetic markers is increased. Apart from the livestock species, sex pre-selection is also important in breeding of companion animals, zoo animals and endangered species.

Earlier, because of the absence of a practical and accurate *in vitro* method to determine the sex of sperm, the only way was to breed animals. After understanding the principle behind sex determination and differences between X and Y specific spermatozoa, many techniques were formulated in order to alter the sex ratio at birth. Advancement in molecular biology and biotechnology improved these techniques. Later, the Beltsville Sperm Sexing Technology became very popular as it applies to a wide range of species. Now a days there are many simple methods which can be applied even at farm level. It is essential to have an outlook on the whole scenario emphasizing the main adaptable techniques at the field level.

Sex Ratio and Levels of Sex Ratio Manipulation

The ratio of males to females in a population is called the sex ratio. It is practically calculated as the number of females per thousand males. In the broad sense, it applies to any species with male and female sexes. The X and Y chromosomes evolved from a pair of identical chromosomes (Muller 1914, B Lahn 1999) termed autosomes, due to an allelic variation in an ancestral animal, and this led to the development of a "sex locus". The chromosome with this allele became the Y chromosome, while the other member of the pair became the X chromosome. Over time, beneficial genes for males and harmful genes for females either developed or translocated on the Y chromosome (Graves, 2006). In mammals, chromosomal sex determination and the assignment of a sexual identity to a child are both determined solely by the gametes, or sperm. A male generates two forms of sperm, known as X-sperm and Y-sperm, whereas a female's oocyte (ovum) will always carry an X-chromosome as it matures into an ovum. Because of this, the fertilisation of an ovum by a Y-sperm results in the production of a male (XY), whereas the fertilisation of an ovum by an X-sperm results in the production of a female (XX). When it comes to mammals, the X-sperm have a greater amount of DNA than the Y-sperm.

Ernst Mayr (1937) has classified the sex ratio into primary, secondary, and tertiary. The primary sex ratio is the ratio of both gametes at fertilization. It can be described as the sex ratio of unborn offspring in a population and is expected to be near 50:50 in natural populations. But there are a few cases where certain factors can change the primary sex ratio by, for example, having lethal effects on sperm. The secondary ratio is the ratio at birth, often nearly 50:50 but research suggests that the secondary sex ratio skews more than the primary sex ratio. The tertiary sex ratio or adult sex ratio (ASR) is the ratio in sexual adulthood where the maximum skewing of the sex ratio is observed mostly observed. Another term operational sex ratio (OSR) is the proportion of adults in the sexually active population that are males. The quaternary sex ratio is the male/female ratio in adults past the age of sexual reproduction. It is of little importance in a livestock population apart from making culling decisions. Differential survival rates can lead to an imbalance between these ratios. In the case of *in vitro* fertilization, the manipulation can be done from the primary sex ratio level. The number of female animals born on a farm has greater significance because it can directly affect the profit not only for dairy animals but also in the case of solely meat-purpose animals like pigs.

Principles Behind Sex Pre-selection

Individual detection of X and Y sperms and separating them precisely without loss of viability offer a chance to preselect the gender of offspring to be born in agriculturally important animals. This idea of sex pre-selection has been practiced and evolving for centuries but mostly the results were not satisfactory. Shortcomings of the earlier practices resulted from the lack of knowledge about the underlying principles behind sex determination and the difference between X and Y chromosomes. Wrong interpretations like right testis produced males while left produced females were prevailing as early as 470 BC, in Greece (Garner et al., 2003). Later the advancement of molecular biology enlightened the scientific community about the genome, sex-determining factors in chromosomes, and the difference between X and Y chromosomes. Then also, sorting semen remained complicated by the fact that X and Y spermatozoa are mostly identical in size, shape, weight, electrical charge, and speed, which is nature's way of providing a 50/50 chance for male or female offspring. Nevertheless, the X chromosome is having 4% more sperms than in Y chromosome (Moruzzi et al., 1979). This can be identified in terms of contrast in fluorescence with the help of a DNA-binding dye, laser light, and computer analysis. Also because of the less DNA content Y-bearing spermatozoa are thought to have quick motion in comparison with X-chromosome-bearing sperms. Sex preselection at the embryo level can be done by detecting Y-specific regions in the chromosome.

Techniques of Sperm Sorting

At first, the techniques such as Albumin gradient, Percoll density gradient, swim up, etc. were attempted for separating the X & Y spermatozoa. Later vital staining using fluorochromes such as acridine orange, ethidium bromide, or mithramycin became popular. Then came bis-benzimidazole dyes Hoechst 33258, Hoechst 33342, and DAPI (Keeler et al., 1983). There are both conventional/traditional methods and molecular methods for sorting stained sperms of the preferred gender.

1. Quinacrine Mustard Staining

It is predicated on the fact that the dye has a very great affinity for the heterochromatic areas of the Y-chromosome, and that these regions glow very brightly when exposed to UV light. However, quinacrine causes both false positive and false negative findings in interphase cells, and multiple investigations have demonstrated that this method might produce results that are deceptive and imprecise. In addition, quinacrine produces false positive and false negative results in interphase cells (Flaherty and Matthews, 1996). In addition, quinacrine fluorescence is not an inherent characteristic of the Y chromosomes of all mammalian species. Therefore, it is considered an inappropriate approach for the selection of sperm for most mammalian species.

2. Centrifugal Counter Current Distribution based on Density Characteristics

This technique was based on the density difference between X and Y bearing Spermatozoa. Centrifugation technique is used to separate spermatozoa into stationary and mobile phase. But later, Meistrich (1982) showed that the density difference between X- and Y-bearing bovine spermatozoa was just 0.0007 g/cm^3, rendering it less suitable for sexing sperm.

3. Sperm Sorting Based on Volumetric Differences

The head volume of the spermatozoa can be used as a criterion to distinguish X- and Y-chromosome-bearing sperm. This also has an advantage over the techniques based on DNA content because it does not need any staining procedure.

4. Gradient Separation of X and Y Sperms

The gradient separation procedures depend on the natural movement of sperm cells in a discontinuous column.

a) **Albumin Gradient:** In 1970, Ericson et al. reported the use of an Albumin gradient for the separation of X and Y-bearing spermatozoa. He claimed the apparentenrichment of Y-sperm was due to their faster movement through the viscous albumin. Later According to research published by Maxwell et al., (1984), this technique successfully increased the percentage of motile spermatozoa and reduced the number of defective spermatozoa. Now a days nano albumin gradient is successfully used for separation (Maulana et al,2022).

b) **Percoll Density Gradient:** Spermatozoa are allowed to pass through a layer of semen that is placed on top of a percoll column that is not continuous. Since in comparison to X sperms Y sperms are lighter, they will not be able to penetrate the denser lower layer even after centrifugation resulting in X sperm enrichment in the lower layer.

5. Swimming Patterns under Laminar Flow

This technique based on the observation that spermatozoa carrying the Y chromosome travel more quickly and differently in a column of flowing media than spermatozoa carrying the X chromosome do.

6. Free flow Electrophoresis

The separation of X-bearing and Y-bearing spermatozoa is accomplished with the use of an electric field, which is based on the idea that there may be a difference in the electric charge that is present on the surface of the two types of spermatozoa (Kaneko et al., 1984). The popularity of this method is, however, hindered by the fact that electrophoresis causes a loss in motility.

7. Counter Current Galvanic Separation

In this method, Y-bearing spermatozoa are drawn to the anode while X-bearing spermatozoa are drawn to the cathode when an appropriate micro-ampere current is supplied (Bhattacharya, 1977).

8. Immunological Sexing of Semen

Immunological sexing of sperm is done by using anti-X and anti-Y antisera raised in male and female rabbits' bodies upon immunization with a combination of sperm preparation and Feund's incomplete adjuvant. This anti-sperm antiserum, when incubated for 60 min at 38.5 °C and 5% CO_2 was showing agglutination of spermatozoa. The population of free-swimming sperm is isolated after being separated from the agglutinated sperm population using glass wool filtration. This process allows the agglutinated sperm population to be removed. However, the rate of its success in a variety of species needs additional research.

Polymerase Chain Reaction (PCR) and Fluorescence *in situ* Hybridization (FISH)

Using PCR and FISH, it is not only possible for accurate detection of X and Y sperms but also for validation of different semen sorting techniques. There have been reports of specific DNA sequences on X- and Y-sperm of a variety of livestock species, and these sequences can be utilised to determine the gender of an individual sperm as well as the gender ratio of sperm in samples of semen. FISH utilizes specific fluorescence-conjugated probes for accurate identification of X and Y-sperms while PCR amplifies specific X and Y regions in the spermatozoa. The quantitative real-time PCR, also known as qPCR, is a method that makes determining the gender ratio of sperm both easier and more accurate (Parati et al., 2006). The evaluation of sex selection procedures, on the other hand, is not helped much by its application to populations of cells. In addition to this, the screening of a large number of different sperm samples requires a significant amount of manpower (Flaherty and Matthews, 1996).

9. Flow-cytometric Sorting of Semen

Flow cytometric sorting of sperms works based on Qualitative flow cytometry. In this technique spermatozoa are labelled with DNA fluorescent dye Hoechst 33342. After co incubation and labelling it is hydrodynamically focussed in a discontinuous droplet stream and UV laser beam is passed through this. The flow cytometric sorter allows the identification and selection of individual spermatozoa with a sorting purity above 90%% (Johnson, 2000). Even though the flow cytometric approach was capable of precisely determining DNA content differences between X- and Y-sperm from cattle, sheep, pigs, and rabbits since the sperm loses viability during the process it was of no use to sort the samples to live X and Y sperms.

Complete Sexing v/s Enrichment

As of now, the most reliable and robust tool for separating X- and Y-chromosome-bearing spermatozoa (CBS) is fluorescence-activated cell sorter (FACS) which utilizes the DNA content differences found in X- and Y-sperm of different mammalian species. Although it has got several advantages, several factors currently limit its utility in farm animals. The low sperm output of a flow sorter and its practical utility at the farm level is the main limitations to the widespread use of this technology. Further, in litter-bearing animals, the low dose of Artificial insemination may cause a decrease in litter size which would counter the advantage expected from the sexing of semen. Also, there are altered mRNA expression patterns of developmentally important genes in *in vitro* produced embryos using sorted semen, showing that sperm sexing by flow cytometry may result in sperm damage, semen fertility reduction, and ongoing pregnancy failure. The high number of stressful steps during the procedure of sex-sorting and freeze-thawing increases the number of dead and damaged spermatozoa, resulting in low in vivo fertilizing ability. These drawbacks limit the use of sophisticated techniques. So, in order to exploit the socioeconomic benefits of getting more female animals, adoption of sperm enrichment techniques which can be practiced at field level has to be done. But the efficiency of these methods must be evaluated in case of each species and suitable ones with better results has to be modified at individual farm level.

Sperm Enrichment -Issues and Prospects

The sorting techniques sometimes affects the morphology or fertilizing capacity of spermatozoa. The DNA specific dye causes chromatin decondensation and application of high pressure or speed may cause DNA damage. Fertilizing ability and viability may be reduced by the application of electric field. The commercially available technology for separating X- and Y-bearing spermatozoa is flow cytometry with 90% accuracy (Johnson, 2000) and the use of sexed semen technology is the apparent lower fertility of sorted sperm (Palma et al., 2008). The sorting procedure might be responsible for variations in cleavage (0 to 89%) and blastocyst rates (3.5 to 28.8%) (Palma et al., 2008; Borchersen and Peacock, 2009). There are altered mRNA expression patterns of developmentally important genes in *in vitro* produced embryos using sorted semen, showing that sperm sexing by flow cytometry may result in sperm damage, semen fertility reduction (Morton et al., 2007), and ongoing pregnancy failure (Perez-Crespo et al., 2008). Requirement of skilled manpower and operational cost also reduces the application of most techniques.

Therefore, these selection methods are modified to sort the X and Y spermatozoa without reducing the viability and fertilizing capacity. Some of these techniques such as the swim-up (Madrid-Burry et al., 2003) method and Percoll™ density gradient centrifugation (Bhatt and Sharma, 2020) are suitable for adaptation at farm conditions as simple procedures at a reduced cost. Successful attempts of sex preselection using field adaptable enrichment techniques in different livestock species are listed below.

Cattle and Buffalo

In cows, Madrid-Bury et al. Madrid- Bury, (2003) showed that a routine swim-up procedure cannot isolate motile sperm fractions with a sex ratio distortion, but their modified swim-up technique using heparin resulted in a high percentage of X chromosome-bearing sperm. Because X sperm have more DNA than Y sperm, the chromosomes that they carry have differing migratory velocities, which leads in the Y-bearing chromosome being quicker than the X-bearing chromosome (Yan et al., 2006). On the basis of this assertion, a number of writers claimed that the supernatant obtained from the swim-up process included a greater number of Y sperm (Check et al., 1994). Researchers Madrid-Bury et al. (2003), Yan et al. (2006), and Cesari et al. (2006) examined the fraction of X- and Y-bearing sperm in the swim-up supernatant of a variety of cattle species and found no significant difference between the two. In their study from 1997, Bottcher-Luiz and colleagues found that the supernatant included 69.6% of X-bearing sperm. Nano albumin gradient was used for successful Separation of X- and Y-Spermatozoa Ongole Crossbred (Maulana et al., 2022). In case of buffaloes, flow cytometric sorting of semen is not much adapted because of the higher cost. Development of a feasible technology is of greater importance in case of buffaloes. Modified swim-up method conducted in Nili Ravi buffaloes did not compromise sperm quality after separation and freeze thawing. Higher proportion of X specific sperms were recovered from lower fraction of the column (Awan et al., 2017). For obtaining Y bearing sperm fraction sucrose density gradient was used in buffaloes (Asma-ul-Husna et al., 2020).

Sheep and Goat

In small ruminants, due to narrow cervical way and different anatomical structure, it is important to ensure the quality of sorted sperm used for insemination. Graaf et al., (2006), conducted experiments sorted ram semen and showed that sex-sorted spermatozoa derived from fresh semen are superior in some respects to those derived from frozen semen. The greatest variation between the DNA content of X and Y spermatozoa can be found in goats, when compared to other livestock species (Parilla et al.,2004). In dairy got industry, large number of female goats are required and there is a benefit of twinning which can increase the profit. Huang et al., (2022) demonstrated that in buck semen X sperm enrichment was observed when swim up method was conducted after incubation at pH 7.4 in alkaline diluent combined with resiquimod (R848). An X -sperm enrichment of more than 85% was attained without change in pregnancy rate. Thirunelvely et al,2013 has done successful separation of X and Y sperms from goats using sucrose density gradient. The fact that Y sperm were enriched in a solution of sucrose with a concentration of 1g/ml demonstrates that this method might potentially be used to establish low-cost sexing of sperm in developing nations.

Pig

Sex pre-selection has major advantages in porcine species since females are preferred for rearing. This can reduce the extra effort for handling, castration and split sex feeding. Moreover, the reluctance to meat of male animals because of boar odor can also be avoided. Usually, two methods are used for sperm enrichment in porcine species. Direct swim-up and percoll density gradient using single or multiple layers. Albumin gradient is also beneficial since albumin promotes sperm capacitation (Gadea et al.,2020). Park et al., (2009) used a modified swim-up method that reduced the usual incidence of polyspermy while doing IVF in pigs and yielded high quality embryos without chromosome abnormalities.

Horse

In the case of equines, especially for rare breeds and for commercial purposes, it is essential to ensure the birth of foals of specific gender. Females are easier to tame but males of good pedigree are used for horse shows. Even though the previous studies on flow cytometric sorting of stallion sperm were successful, the viability of sperms was compromised. Equine sperm sorting was successfully conducted using density gradient separation and a modified freezing protocol by Heer et al., (2007). They were able to get a higher percentage of morphologically normal live equine spermatozoa. But the technique is yet to be standardized for an improved percentage.

Conclusion

Sex pre-selection can accelerate the socio-economic benefits of a livestock farmer. Semen sorting techniques has evolved greatly from time to time. Most of the available techniques are a bit complicated and has a risk of damaging the gametes.This implies the adaptation of sperm enrichment techniques suitable for different species at farm level. Techniques like swim-up, percoll density gradient etc. with modifications demanded by each species has been proved to be successful particularly for X sperm enrichment in different species. These techniques can be standardized at different farm levels with slight effort to ensure the birth of livestock of desired gender.

References

Asma-ul-Husna, A.A., Awan, M.K., Sami, Q.S., Ansari, B.A.R. and Naqvi, S.A., 2020. Efficiency of sucrose density gradient method for sex separation of buffalo spermatozoa as validated by SYBR Green Real Time PCR. Animal Science Papers and Reports, 38(1), pp.21-33.

Awan, M.A., Mehmood, A., Sultana, T., Shahzad, Q., Ansari, M.S., Rakha, B.A., Naqvi, S.S. and Akhter, S., 2017. Sperm sexing in Nili-Ravi buffalo through modified swim up: Validation using SYBR® green real-time PCR. Animal reproduction science, 182, pp.69-76.

Bhat, Y. and Sharma, M., 2020. X-sperm enrichment of bovine semen by percoll density gradient method and its effect on semen quality, sex ratio and conception rate. Indian Journal of Animal Research, 54(10), pp.1181-1187.

Bhattacharya, B.C., 1977. Guide to pre-arranging the sex of offspring. Guide to pre-arranging the sex of offspring.

Borchersen, S. and Peacock, M. 2009. Danish AI field data with sexed semen. Theriogenology. 71(1):59-63.

Bottcher-Luiz, F., Magna, L. A. and Fazano, F. A. 1997. Seleção sexual emdoisprocessos de capacitaçãoespermática: Percoll e Swim-up. Reprod. Clim. 12(3):141-4.

Cesari, A. G. G., Kaiser, N. and Mucci, A. 2006. Integrated morpho physiological assessment of two methods for sperm selection in bovine embryo production in vitro. Theriogenology. 66:1185-1193.

Check, J. H., Kwirenk, D., Katsoff, D., Press, M., Breen, E. and Baker, A. 1994. Male: female sex ratio in births resulting from IVF according to swim-up versus Percoll preparation of inseminated sperm. J. Androl. 33(1):63-65

De Graaf, S.P., Evans, G., Maxwell, W.M.C. and O'Brien, J.K., 2006. In vitro characteristics of fresh and frozen–thawed ram spermatozoa after sex sorting and re-freezing. Reproduction, Fertility and Development, 18(8), pp.867-874.

Ernst Mayr (1937) Mayr, E., 1939. The sex ratio in wild birds. Am. Nat. 73(745):156-179.

Flaherty, S. P., Payne, D. and Matthews, C. D. 1998. Fertilization failures and abnormal fertilization after intracytoplasmic sperm injection. Hum. Reprod. 13(1):155-164.

Gadea, J., Coy, P., Matás, C., Romar, R. and Cánovas, S., 2020. Reproductive technologies in swine. In Reproductive Technologies in Animals (pp. 67-79). Academic Press.

Garner, D. L. and Seidel Jr, G. E. 2003. Past, present and future perspectives on sexing sperm. J. Anim. Sci. 83(3):375-384.

Graves, J.A.M., 2006. Sex chromosome specialization and degeneration in mammals. Cell, 124(5), pp.901-914.

Heer P, 2007: Adjustment of cryopreservation of stallion spermatozoa to the Beltsville Sperm Sexing Technology. Doctoral Thesis, University of Veterinary Medicine Hannover, Hannover, Germany (in German).

Huang, M., Cao, X.Y., He, Q.F., Yang, H.W., Chen, Y.Z., Zhao, J.L., Ma, H.W., Kang, J., Liu, J. and Quang, F.S., 2022. Alkaline semen diluent combined with R848 for separation and enrichment of dairy goat X-sperm. Journal of Dairy Science, 105(12), pp.10020-10032.

Jonhson, L. A. 2000. Sexing mammalian sperm for production offspring: the state-of-theart. Anim. Reprod. Sci. 60:93-107.

Kaneko S, Yamaguchi J, Kobayashi T, Iizuka R (1983): Separation of human X- and Y-bearing sperm using Percoll density gradient centrifugation. FertilSteril 40: 661– 665.

Keeler KL, Mackenzie NM and Dresser DW (1983) Flow microfluorometric analysis of living spermatozoa stained with Hoechst 33342 Journal of Reproduction and Fertility 68 205-212

Lahn, B.T. and Page, D.C., 1999. Four evolutionary strata on the human X chromosome. Science, 286(5441), pp.964-967.

Madrid-Bury, N., Fernández, R., Jiménez, A., Pérez-Garnelo, S., Moreira, P. N., Pintado,B., De La Fuente, J. and Gutiérrez-Adán, A. 2003. Effect of ejaculate, bull, and a double swim-up sperm processing method on sperm sex ratio. Zygote. 11(3):229-235.

Maulana, T., Ridwan, R., Gunawan, M., Agung, P.P., Afiati, F., Kaiin, E.M. and Said, S., 2022. ASuccessful Separation of X-and Y-Spermatozoa Ongole Crossbreed Using a Nano-Albumen Gradient Column. Tropical Animal Science Journal, 45(4), pp.397-404.

Maxwell, W. M. C. and Watson, P. F. 1996. Recent progress in the preservation of ram semen. Animal Reproduction Sci.. 42(1-4):55-65.

Meistrich, M. L. and Brown, C. C. 1983. Estimation of the increased risk of human infertility from alterations in semen characteristics. Fertil. Steril.. 40(2):220-230.

Moruzzi, J.F. 1979. Selecting a mammalian species for the separation of X-and Y-chromosomebearing spermatozoa. Reproduction. 57(2):319-323.

Muller, H.J., 1914. A gene for the fourth chromosome of Drosophila. Journal of Experimental Zoology, 17(3), pp.325-336.

Palma, G. A., Olivier, N. S., Neumüller, C. H. and Sinowatz, F. 2008. Effects of Sex sorted Spermatozoa on the Efficiency of in vitro Fertilization and Ultrastructure of in vitro Produced Bovine Blastocysts. Anatomia, histologia, embryologia. 37(1):67-73.

Parati, K., Bongioni, G., Aleandri, R. and Galli, A. 2006. Sex ratio determination in bovine semen: A new approach by quantitative real time PCR. Theriogenology 66(9):2202-2209.

Parrilla I, Vazquez JM, Roca J, Martinez EA (2004). Flowcytometry identification of X- and Y-chromosome bearinggoat sperm. Reprod. Dom. Anim. 39: 58–60

Park, C.H., Lee, S.G., Choi, D.H. and Lee, C.K., 2009. A modified swim-up method reduces polyspermy during in vitro fertilization of porcine oocytes. Animal reproduction science, 115(1-4), pp.169-181.

Pérez Crespo, M., Pintado, B. and Gutiérrez Adán, A. 2008. Scrotal heat stress effects on sperm viability, sperm DNA integrity, and the offspring sex ratio in mice. Molecular Reproduction and Development: Gamete Res. 75(1):40-47.

Thirunelvely, J. and Lanka, S., Sex Pre-selection by Quantification of Y-Chromosome Bearing Spermatozoa in Goat Species. International Journal of Scientific and Research Publications, p.415.

Yan, J., Feng, H. L., Chen, Z. J., Hu, J., Gao, X. and Qin, Y. 2006. Influence of swim-up time on the ratio of X-and Y-bearing spermatozoa. European Journal of Obstetrics &Gynecology and Reproductive Biology. 129(2):150-154.

12

High Throughput Phenotyping and Big Data Analytics for Improvement of Farm Animals

M. K. Verma and J. Singh

Department of Animal Genetics & Breeding, College of Veterinay Science and Animal Husbandry, Acharya Narendra Deva University of Agriculture and Technology, Kumarganj, Ayodhya, Uttar Pradesh

Introduction

Raising milk and meat production are long-standing global issues. Researchers are learning that present breeding efforts won't be sufficient to meet anticipated future food demands given the current climate scenario. The animals that influence the overall yield of the field must be screened as soon as possible. In order to build more successful modern breeding programmes, high-throughput phenotyping is currently being included into conventional breeding operations. Despite the public accessibility of their genetic information online, there is still a lack of phenotypic data on the genomes of various animal breeds because environmental factors make it more difficult to define phenotypes and increase the likelihood of measurement error (Rahaman et al., 2015). Genomic selection (GS) is a strategy that predicts the genomic estimated breeding values of lines in a breeding population using information from genome-wide markers (Meuwissen et al., 2001). When compared to other conventional methods like marker-assisted selection (MAS), genomic selection has some inherent advantages. These include boosting genetic gain by reducing breeding cycles and acquiring minor effect loci based on markers dispersed throughout the entire target genome. Reliable phenotypes and genotyping are both necessary for the training of accurate prediction models for genomic selection. Phenotyping is regarded as a significant factor restricting genetic improvements in animal breeding due to high labour and time expenditures. Traditional approaches of phenotyping are prohibitively expensive, time-consuming, and slow. They are also typically hazardous, and they can only test a small number of factors at the same time (Hein et al., 2021). As a response, efforts have been undertaken to build high-throughput phenotyping (HTP) technologies.

For effective animal breeding programmes and for the genetic deconstruction of complex traits, accurate and comprehensive phenotypic data are essential. Non-destructive phenotyping, a contemporary technology, adds a new dimension to data gathering by accelerating the accuracy and speed of data processing (Yang *et al.,* 2020). Digital image analysis can be used to extract useful information from pictures and as an input for imaging processing methods that directly relate to phenotyping livestock. Other technological developments could lead to a wide variety of novel phenotypes, including automatic feeding mechanisms, activity monitor devices, and indirect biomarkers at the physiological and cellular levels. Additionally, infrared spectrometry is becoming more and more common in precision livestock husbandry as a useful resource for online analysis and a non-destructive measurement instrument.

Animal farming has always been decentralized and done on a scale that a small number of individuals can control. Additionally, most livestock farmers did not have access to cutting-edge innovations like high-speed internet, smartphones, and affordable processing power until approximately a century ago. Both of these situations are now changing quickly. HTP is necessary because over the next three decades, the need for different meat and other animal products is anticipated to increase by more than 70% on a global scale. Secondly, smart phones or PCs are directly connected to the internet by more than half of the global population.

Presently, a lot of farmers may easily access computational power. We now understand that wherever populations and affluence have grown, so has the consumption of meat. Because there is a shortage of land and other natural resources, we must find more efficient techniques to produce more animals per hectare. We must also create animals using fewer resources, such as less water land, and other natural resources. This suggests that outdated manual animal rearing techniques might no longer be effective. Additionally, it means that in order to increase our revenues from animal husbandry, we must develop strategies and frameworks. Different industries are already being transformed by computers, sensors and some artificial intelligence. The earnings and efficiency they produce are higher (Wolfert *et al.,* 2017). For this reason, we must look into how cutting-edge technologies might help us increase the efficiencies and benefits of animal farming. Large-scale phenotyping is a better method for correct selection in a commercial production system since it allows for the evaluation of genetic features for attributes associated to animal wellbeing. The primary bioinformatic and statistical tools for this aim were discussed, with an emphasis on methods for creating novel HTP-based wellbeing indicator features, such as movement recording with wearable sensors and accelerometers. Big data science in conjunction with sensor-based phenotypes, such as Inertial Measurement Unit (IMU)-based movement capture, infrared thermography, and sound analysis, are essential for translating animal welfare indicator traits into exact genomic breeding values. This can be applied in industrial selective breeding to keep enhancing animal toughness. The beef, dairy, poultry, and swine sectors saw a radical change as a result of the ability to gather and use on-farm data for breeding purposes, leading to enormous productivity and efficiency benefits (Hill, 2016). Using sensors and other high-throughput phenotyping technologies, similar opportunities for huge profits may exist in several other instances of the livestock industry.

Sources of Important Expenses in Animal Farming

In animal farming, the stocking rate, food, and disease control are all potential key cost factors. Because of economies of scale, farmers can reduce their production costs and optimize their key expenses by adding more animals to a system. The cost of the feed, which makes up more than 40% of the costs associated with producing milk, has the greatest financial influence on the profitability of dairy farms. By reducing greenhouse gas emissions from livestock and manure, as well as the amount of land needed for manure disposal and water requirements, feed efficiency improvements will also assist the environment. An epidemic of an infectious disease might cause enormous losses in a big farm housing thousands of animals. In such a situation, the outbreak of an infectious disease will be difficult to control unless the farmer makes prompt early treatments. Often, it is too late to take action once symptoms start to develop. Uncontrolled disease spread leads to animal fatalities, worsened health outcomes, and economic losses.

High-Throughput Phenotyping (HTP) Platforms

Numerous high-throughput phenotyping techniques and platforms have been applied to animal studies over the past ten years in order to: identify functional limiting factors; determine the ideal nutrient composition of animal feed; assess animal management in order to assess performance; investigate techniques for reducing dietary excretion into the surroundings; or forecast results in other covariates. The present non destructive high-throughput phenotyping platform tracks the position, speed, and duration of movement using advanced technologies including activity sensors like accelerometers pedometers, and GPS devices. The application of non-invasive sensors can enable the phenotyping of thousands of animals in a single day, which is analogous to the impact that high-throughput DNA sequencing technology has had in animal genomics (Finkel et al., 2009).

Phenotyping Platforms in Trait

Many problems in animal farming can be overcome by utilizing cutting-edge technologies. Examples include identifying the best ways to lower costs, increase output, improve efficiency, and create the finest diet formulations. In order to offer pertinent and contextually optimum answers, advanced models may even take into consideration factors like genetics, environment, and managerial priorities. In general, a system's odds of producing correct and ideal answers increase with the variety of datasets it collects and analyses (Ellis *et al.,* 2020).

Detecting Animal Health

Important animal health factors like mobility, air quality, and food and hydration consumption have all been continuously monitored using high-throughput techniques. Farmers may now identify, predict, and prevent disease outbreaks even before a large attack by continuously collecting this data and using cutting-edge Artificial Intelligence and Maximum Likelihood algorithms to foresee deviations or anomalies (Neethirajan *et al.,* 2020). This kind of approach can lower the cost of production while also warning farmers about the potential for disease even in the preclinical stage. As a result, farmers will be better able to act quickly to prevent catastrophic losses. They can quickly predict and halt the spread of infectious diseases like the African swine flu for a lot less money.

Prediction of Lameness in Early Ages

In some cases, algorithms can predict an animal's propensity for an illness based on its movements, such as lameness. Modifications in gait, excessive use of some body parts, and inactivity in other body parts are all reliable indicators of preclinical lameness (Taneja *et al.,* 2020). Due to its negative effects on milk production and increased risk of damage, lameness is the third most significant illness affecting agriculture. Farmers can prevent major financial losses by anticipating lameness. According to the findings of another study, infected pigs walk around up to ten percent less during the first two days following infection. This can be used to isolate diseased animals before they spread the disease to numerous other animals.

Detecting Animals in an Extensive Production System

Animals in a large system have been located using HTP, and data about cattle in large production systems can be obtained using remotely sensed imagery (very high-resolution satellite imagery, aerial photography and unmanned aerial vehicles), as well as Convolutional Neural Networks (CNNs) as Artificial Intelligence technology for object-based detection of cattle in large datasets. Utilizing the footage captured by unmanned aerial vehicles (UAVs) to identify certain animals and their locations is one way to acquire information on the robustness and efficiency of an operation.

Observing the Lying Behavior of Farm Animals

Ultra wide band (uwb) positioning systems can be used to study an animal's laying habits. It has been shown to alter in response to alterations in welfare and health status (Tucker *et al.,* 2021). For instance, lameness lengthens overall laying time while decreasing the number of times an animal gets up or lies down. Similar to how metabolic problems that affect ruminating duration and udder infections that make an animal very unwell will affect laying behaviour. Reliably detecting and tracking lying over time can provide information on welfare and health state, help develop new precision phenotypes, and accurately and non-intrusively assess things like management styles or living conditions. As the number of animals maintained per farm labour unit increases, it becomes harder to visually identify lame cows in larger herds. Typically, lame cows are detected by their stride and back posture.

Monitoring Resilience in Animals

Dairy cows that are considered lifetime resilient are those who have a high possibility of finishing several lactations, good reproductive and productive performance, minimal serious health issues, and efficient and reliable milk supply (Adriaens *et al.,* 2020). Enhancing dairy cow resilience has many advantages, including positive effects on farm productivity, the need for antibiotics, and the environmental impact of the industry. The ability to combine continuous data recordings from other sensors to improve resilience prediction has recently become possible thanks to technology developments and an increase in the use of other sensors.

Forecast to Animal Survival

Cow survival is a complex feature that takes into account attributes including milk output, fertility, health, and environmental aspects like farm management. Another indication of effective farming practises in terms of animal care is a high farm average of lactations attained. Farmers would be prudent to just keep those cows that are likely to survive in a production environment because there are many benefits to cows that live long, productive lives. It would be able to choose cows with a high likelihood of surviving to higher lactations by forecasting a cow's capacity for survival at an early stage. The aggregation approach, also known as the ensemble method (Wozniak *et al.,* 2014) or hybrid classifier (Wozniak *et al.,* 2014) can be employed to predict survival in farm animals (Satopaa *et al.,* 2014).

Enhancing the Swine Farming

Computer vision technology has been applied to a number of pig farms. The most significant approaches to our issue are those that make *in vivo* analyses or estimations of carcass composition utilising image-based techniques (Carabs *et al.,* 2016). Previous pig farming applications have included automated weight calculation, walking pattern analysis, and monitoring and identifying pig activities in pens. Pigs can be grouped at the start of the finishing phase by using the ML algorithm to predict their slaughter age. Which ejaculate should be processed for AI dosages can be determined using Gradient Boosting Machine (GBM) models. Pigs that are prone to abnormal growth rates, meat percentages, or pneumonia during the growing-finishing phase can be recognized using ML algorithms like the Gradient Boosting Machine algorithm; early detection would assist in avoiding heterogeneous pens through management.

Management of the Calving in Dairy Farm

Dairy cows and their calves wellbeing and survival depend on proper calving management. Farmers might examine cows for calving more effectively if they used a sensor system that anticipates the moment of calving. It's crucial to keep an eye on a cow before she gives birth because dystocia, which can have detrimental repercussions on both the cow and the calf, can happen. Sensor data can be more useful in predicting the beginning of calving than the anticipated calving date alone. Several behavioural and physiological characteristics related to the start of calving can be automatically monitored by sensors. The final two weeks before calving, dairy cows' eating and rumination activity steadily declines, and after calving, it drastically lowers, sensors appear to recognize able to recognise these differences. Dry matter intake and feeding frequency both decline in the 24 hours prior to calving as well as activity levels, an activity index may reliably predict whether a cow would calve within 6 hours of an increase in the index. Temperatures of vulva, rectum, and rumen were decreases over the 24 hours before calving. All of these sensor-measured characteristics have predictive value for calving (Ouellet *et al.,* 2016). Farmers could determine when a cow needs close monitoring if the onset of calving could be predicted more precisely than the anticipated calving date.

Oestrus Detection

Oestrus detection is one of the most time-consuming, complex, and error-prone activities that dairy farm workers must perform. Poor oestrus detection performance can be attributed to a variety of reasons, including inexperienced farm workers, ineffective use of aids, and weak oestrus symptoms by dairy cows (Burke *et al.,* 2012). One such commercially accessible solution that makes use of pedometer or accelerometer technologies is activity-based oestrus detection systems (ODS). These systems are predicated on the idea that an increase in locomotor activity is linked to oestrus.

Development of the Biomarkers

Deep phenotyping can provide biomarkers that are sensitive, specific, and reasonably inexpensive for identifying the trait or disease phenotype of interest. This can help ensure that animals are classified in the appropriate categories. A biomarker is any substance or process that can be detected in a biological specimen and that is inextricably linked to the feature of interest. Biomarkers can be used to study everything from diseases to cancer to infectious diseases. In addition to being based on other modalities like imaging, biomarkers can also be molecules like RNA, metabolites, microorganisms, or proteins/peptides (Hartwell *et al.,* 2006). A minimally invasive biomarker assay is one that can be found in peripheral blood or urine. Screening tests based on biomarkers are also used for diseases that have not yet reached the clinical stage, are asymptomatic, or are in their early stages. In addition, the benefits of early intervention and prevention need to have a greater impact than the expenses of the screening test.

Big Data

The cattle sector understands that there are potential and difficulties to be more effective, environmentally friendly, and socially responsible in order to fulfil the requirements of a growing human population. Opportunities for more lucrative and sustainable agriculture are presented by these difficulties. It is feasible to make a link between the phenotype and the genotype, construct new sensor technologies, develop new data coordination systems, and create computational tools for analyzing large amounts of data, and ultimately use this knowledge to improve animal production and welfare thanks to the variety of data types needed to address these grand challenges (Panigrahi *et al.,* 2020). The amount of data gathered from individuals and groups of animals has expanded tremendously as a result of the accessibility of cheap processing power, large storage devices, and internet connectivity. When the appropriate analytical framework is applied, combining records of individual animals, such as weights, treatments, and characteristics of the carcass, with cohort-level data, such as daily feed deliveries, diet components, group weights, and movements, results in the production of large amounts of data that are useful for analysis. These data can be useful for analysis only when the appropriate analytical framework is applied. The availability of data on a wide range of hosts, diseases, and environmental factors impacting animal health is increasing exponentially as automated technologies for high-input data collecting become increasingly prevalent. Additionally, sensors, drones, and satellites frequently gather enormous volumes of data. Sensor technologies are being utilised more and more to monitor the health and welfare of cattle, pigs, and poultry. The era of Big Data, in which data sets are of lower accuracy and are gathered fast and in big amounts, and in which equipment may vary widely and in less coordinated ways, is being ushered in by this transition, which is advancing animal research towards (Patra *et al.,* 2021). Big data are defined in many different ways by reliable sources (commercial technology companies like Microsoft and Google are widely referenced sources of description), but a common thread among these definitions is the "Vs" of big data. The so-called four V model defines big data by its collecting characteristics. (IBM, http://www.ibmbigdatahub.com/infographic/four-Vs-big-data), however a literature study reveals that the Vs are growing at the same rate as the data itself, from 4- to 7- to 10- to 42-V models. The four V models use four fundamental characteristics to identify big data: volume, velocity, variety, and veracity. The goal of precision agriculture is to use regular, in-depth observations of individual animals to make the best management decisions possible. A livestock environment might very well mandate the clustering of animals with comparable nutritional or other management demands, the identification of changes in production, the evaluation of reproductive status, the early identification of health issues at both the individual and group levels, and other similar activities.

Big Data Analytics

The practise of studying vast, varied data sets using cutting-edge analytical methods is known as big data analytics. These varied data sets range in size from terabytes to zettabytes and contain unstructured, semi-structured, and structured data from many different sources. For instance, recursive numeric positions of specific animals are recorded by accelerometers and real-time location systems in comparatively brief time increments. These motions can be compared to expected ranges using this data, and algorithms can utilise it to spot patterns linked to physiologic changes like estrus or disease. When paired with individual animal activity monitoring devices, cohort and individual animal data can produce sizable data sets that can be used for both descriptive and predictive analytics. The process of using scientific knowledge to make decisions consists of developing a hypothesis or prediction, putting it to the test, evaluating the results, interpreting the results, revising the procedure, and then repeating it (Larson and White, 2015). A growing library of potential machine learning algorithms is accessible for testing, and classification algorithms vary in their methodologies or strategies for minimizing variation in outcomes of interest. Certain results or data types suit some models more than others. Different models can be tested, and the optimal method can be chosen based on classification accuracy as opposed to statistical modelling, where a single model is selected before analysis. The process of predictive analytics consists of a number of processes, the most important of which are selecting a target variable, processing the data, splitting the data, developing algorithms, refining algorithms, and, lastly, testing the accuracy of the classifiers that have been produced. Each stage of the procedure is essential to guaranteeing that the results are internally consistent and supply the data required to enhance subsequent decision-making. Predictive analysis models enhance farm efficiency and promote good outcomes in animal farming by utilizing data, statistics, and machine learning techniques.

Determine the Problem and the Data Collection

Finding the target variable, which is the result of the data that needs to be estimated or forecasted, is one of the first steps in the predictive analytics process. This is the result of the data that needs to be estimated or predicted. When choosing the objective variable, care should be taken to ensure that it will generate data that will have an impact on significant business decisions that are necessary for the long-term financial stability of the operation. The structure of the predictive analytics problem and the model that is most appropriate for solving it are both decided by the type of data that defines the target variable. When the target variable has been decided upon, the question that has to be answered should be defined in a very specific way.

Preprocessing of Data

The data must be preprocessed after the target variable has been identified and the data structure has been assessed using simple data summarization. The approach often involves identifying probable outliers, estimating the amount of missing data, and, if necessary, utilizing techniques for measuring co-linearity between variables (crucial if using traditional methods like linear regression). When it comes to evaluating production data acquired from livestock systems, many machine learning algorithms handle missing data in a satisfactory manner. This is one advantage of employing machine learning algorithms as opposed to conventional statistical methods. Having high amounts of missing data, on the other hand, might cause variables to become inaccurate predictors; hence, the amount of missing data should be minimized whenever it is practicable to do so. Before beginning the modelling process, you should give some thought to what caused the missing data.

Partitioning of Data

The variables that will be part of the model are finished, and then training and validation/testing data sets are created using the available data. In order to assess the performance of predictive models and make sure that the model can be applied to new datasets in a reasonable manner, data must be divided into various data subsets prior to model creation. The initial subset of data is training data, which contains around (80). The training data subset is utilized to create the initial models, and the frequency of occurrence of the results may vary from the original data.

Model Building

There are a wide variety of various types of predictive models that may be used to a particular data collection in order to classify the data or generate predictions based on the data. Due to the large number of variables that big data contains, many of which interact with one another and have intricate relationships with the desired result, it is frequently impossible to select a preexisting model based on the goal variable. By comparing the performance of various models, the evaluation avoids making assumptions about the manner in which the data will be presented, and the best model is chosen based on final accuracy rating rather than preconceived expectations. The environment for testing and comparing various classifying algorithms to find the best match (in terms of accuracy) for a given event and target variable is provided by the predictive analytic framework.When it comes to handling missing data, dealing with various attribute types (continuous or categorical), generalizability overall, and the capacity to clearly explain why a given prediction was made, classification algorithms vary.

To build Predictive Model

The next step after preprocessing is to train predictive models. The target variable is defined, the necessary parameters are provided to each algorithm, and the model is trained using the partitioned training data subset. Rapid Miner (Rapidminer Inc., https://rapidminer.com, Boston, MA), Knime (Knime Analytics Platform, https://www.knime.com, Zurich Switzerland), and R (R Core Team, http://www.R-project.org/, Vienna, Austria) All of them feature packages or nodes that offer a variety of models that may be tested for categorization issues. A prediction model is developed using the subset of training data, and it is then applied to classify the revising data. The classification accuracy of the revised data is used to assess the preliminary findings.

Validation

To assure the precision of the model's outputs and make it usable in real-world applications, it is not enough to simply create a machine learning model and rely on its predictions; you must also examine and evaluate the model's correctness. The accuracy and biases of the validation process must be ensured by selecting the proper validation method. For instance, if the data amount is sufficient to represent the complete population, validation might not be necessary. Since the sample or training data sets we are using might not precisely represent the population, the scenario is different in the real world. Validation guarantees that the statistics of our new datasets still match, looks for data abnormalities, and that the data schema hasn't changed.

Deployment and Model Monitoring

Model deployment is the process of integrating a machine learning model into a current production environment so that it can accept input and output. Making predictions from a trained ML model accessible to others, such as users, management, or other systems, is the aim of model deployment. The architecture of ML systems, or the organisation and interactions of software components within a system to achieve a stated goal, is closely related to model deployment (Opeyemi *et al.,* 2019) Monitoring has many uses, including early instability detection, understanding how and why model performance degrades, and identifying specific failure scenarios.

Present Challenges in High Throughput Phenotyping and Big Data Analytics

There are currently very few commercial sensors available for continuous automatic real-time monitoring and reliable disease management in cattle production. For instance, there are no sensors available to analyse biomarkers from the breathing areas of cows and pigs, which could reveal the gut flora or even the metabolic conditions of the animals. This gap calls for the creation of sensors and biosensing systems that, among other things, assess biomarkers, miRNAs, and other volatile metabolites using "omics" and non-omics methodologies. There are also certain technological difficulties, like the placement of the sensor, the sample rate, and the transmission method.

Choosing Machine Learning Algorithms

Determining the types and quantities of ML characteristics needed, as well as the optimal algorithms to solve the classification problem, can affect the desired outcome of the animal welfare study. For instance, only five to seven features—out of a total of 44—might be needed to generate results that are extremely precise. Large feature sets in real-time systems may therefore provide challenges due to increasing computational complexity and storage needs. One of the biggest technical obstacles to long-term and real-time monitoring of farm animal behaviour, besides energy issues, is "concept drifts." When a sensor platform and data analysis system must adjust to a change in data distributions inside a concept, concept drift occurs.In supervised classification issues, it is typical to assume that the design model's data was randomly selected from the same distribution as the future classification targets. This is an untenable assumption given the dynamic nature of many different categorization problems.

Data Sharing

Big data in livestock faces specific issues, such as how to share data between institutions and commercial organizations and how to standardize data collection, management, quality assurance, trait terminology, and data processing. To separate information from noise in data, create precise prediction models, and integrate data from various sources and places, creative ways will be needed.

Search of More Complex and Better Results

The era of sensors, big data, and machine learning is currently in effect. Over the coming ten years, it is anticipated that these cutting-edge technology would increase animal farming's productivity and yield. Additionally, it will lessen human mistake. Productivity, farmer earnings, and animal welfare will all rise as a result. In addition, technology may help us improve animal welfare, which has the potential to go beyond increased revenues and production. It might also help with the development of more compassionate, humane, and environmentally sound methods.

Conclusion

In conclusion, agricultural HTP technology has the ability to reduce breeding cycles, improve genetic variation detection, and increase selection intensity and precision to solve the breeder's equation for maximum genetic gain.

References

Adriaens I, Friggens NC, Ouweltjes W, Scott H, Aernouts B, Statham J. 2020. Productive life span and resilience rank can be predicted from on-farm first-parity sensor time series but not using a common equation across farms. Journal of Dairy Science. 103(8):7155-7171.

Burke CR, Meir S, Pellow R. 2012. Use of progesterone profiling to investigate factors affecting conception rates in a large, pasture-grazed dairy herd. Proceedings of the New Zealand Society of Animal Production. 72:45- 50.

Carabús A, Gispert M, Font-i-Furnols M. 2016. Imaging technologies to study the composition of live pigs: A review. Revista de Investigacion Agraria [Spanish Journal of Agricultural Research]. 14(3):e06R01.

Finkel E. c2009. With 'phenomics,'plant scientists hope to shift breeding into overdrive.

Hartwell L, Mankoff D, Paulovich A, Ramsey S, Swisher E. 2006. Cancer biomarkers: a systems approach. Nature Biotechnology. 24(8):905-908.

Hein NT, Ciampitti IA, Jagadish KS (2021) Bottlenecks and opportunities in fieldbased high-throughput phenotyping for heat and drought stress. J Exp Botany

Hill WG. 2016. Is continued genetic improvement of livestock sustainable? Genetics. 202(3):877-881.

Larson RL, White BJ. 2015. First steps when using scientific literature in clinical veterinary practice. J Am. Vet. Med. Assoc. 247:254-258.

Meuwissen T. H. E., Hayes B. J., Goddard M. E. (2001). Prediction of total genetic value using genome-wide dense marker maps. Genetics, 157 1819–1829.

Neethirajan S. 2020. The role of sensors, big data and machine learning in modern animal farming. Sensing and BioSensing Research. 29(100367):100367.

Opeyemi B. 2019. Deployment of Machine learning Models Demystified (Part 1). Towards Data Science.

Ouellet V, Vasseur E, Heuwieser W, Burfeind O, Maldague X, Charbonneau É. 2016. Evaluation of calving indicators measured by automated monitoring devices to predict the onset of calving in Holstein dairy cows. Journal of Dairy Science. 99(2):1539-1548.

Panigrahi M, Kumar H, Sah V, Dillipkumar Verma A, Bhushan B, Parida S. 2020. Transcriptome profiling of buffalo endometrium reveals molecular signature distinct to early pregnancy. Gene. 743:144614.

Patra B, Panigrahi M, Kumar H, Kaisa K, Dutt T, Bhushan B. 2021. Molecular and phylogenetic analysis of MHC class I exons 7-8 in a variety of cattle and buffalo breeds. Anim. Biotechnol, p. 1-7. Advance online publication.

Rahaman M. M., Chen D., Gillani Z., Klukas C., Chen M. (2015). Advanced phenotyping and phenotype data analysis for the study of plant growth and development. Front. Plant Sci. 6:619

Satopää VA, Baron J, Foster DP, Mellers BA, Tetlock PE, Ungar LH. 2014. Combining multiple probability predictions using a simple logit model. International Journal of Forecasting. 30(2):344-356.

Taneja M, Byabazaire J, Jalodia N, Davy A, Olariu C, Malone P. 2020. Machine learning based fog computing assisted data-driven approach for early lameness detection in dairy cattle. Computers and Electronics in Agriculture. 171(105286):105286.

Tucker CB, Jensen MB, de Passillé AM, Hänninen L, Rushen J. 2021. Invited review: Lying time and the welfare of dairy cows. Journal of Dairy Science. 104(1):20-46.

Wolfert S., Ge L., Verdouw C., Bogaardt M.J. 2017. Big data in smart farming—A review. Agric. Syst.153:69–80.

Woźniak M, Graña M, Corchado E. 2014. A survey of multiple classifier systems as hybrid systems. An International Journal on Information Fusion. 16:3-17.
Yang W, Feng H, Zhang X et al (2020) Crop phenomics and high-throughput phenotyping: past decades, current challenges, and future perspectives. Molecular Plant 13(2):187–214

13

Integrating Phenomics and Genomics to Analyze the Body Condition Score in Cattle

***Sanjeev Singh*[1] *and Aishwarya Dash*[2]**

[1]*ICAR-National Bureau of Animal Genetic Resources, Karnal Haryana*
[2]*Animal Genetics and Breeding Division, ICAR-National Dairy Research Institute Karnal, Haryana*

Energy is essential for each and every single biological reaction in the body of an individual. Production, reproduction and physiological functions of an animal depend on its body state and utilization of energy from dietary source and its body reserves. Energy reserves are determined primarily by the relative amount of fat accessible as an energy source and are indicative of health condition. Adipose tissue (AT) is a primary reservoir of energy and recognized as an immunologically and endocrinologically active tissue. Cattle have the ability to utilize body reserve to meet energy demand due to negative energy balance during peripartum period, which is crucial for maintaining their health lactation and neonates. There are several methods to measure tissue reserves like body weight, metabolic and hormonal factors, respiratory calorimeter body water by dilution with D_2O, estimation of fat cells diameter, ultrasound techniques etc. Body weight is not a good indicator as it is affected by different factors such as; age, parity, gestation, body frame, gastrointestinal contents weight of vital organs and shows a greater variation in measuring fat stores. Other methods are inapplicable in field conditions as they are costly, time consuming and need laboratory facilities and skilled technicians. However Body Condition Score (BCS) is an easier and the most practical method, which is an assessment of thickness of fat cover, prominence of bone of tail and head region.

Body Condition Score

Body condition is a good reflection of health status of a herd and performance of economic traits. Murray (1919) explained body condition as the ratio of body fat to non-fat components in live animal. Direct measurement of AT mass (fat) is based on slaughter studies and carcass composition. BCS is an indirect, but inexpensive and rapid method to estimate fat depots and rate of lipolysis in different stage of lactation (Siachos et al., 2021). BCS is defined as a subject scoring method to estimate body fat reserves mainly subcutaneous fat in caudal and dorsal regions (Wight and Russel, 1984; Edmonson et al., 1989). It is universally accepted, non-invasive, quick and inexpensive method to estimate the degree of fatness. Change in BCS overtime is determined by changes in feed intake, utilization of energy for yield, growth and maintenance and in body tissue deposition and mobilization. . It is a method for determining the relative fatness of beef cattle.

Importance of BCS

BCS is an important indicator as it shows an immediate appraisal of body state of animal or health status of herd. Therefore, disease proneness and health potency of cattle can be evaluated. BCS of animal gives an inference about its genetic potential in terms of feed utilization and conversion efficiency. Therefore, it may help in customization of feeding strategies of animals at different stages of lactation. BCS is associated with reproductive efficiency such as, shorter calving interval, post-calving estrus recovery, rebreeding performance in subsequent breeding seasons etc. So, it can be used as an indicator for fertility traits which have low heritability and are difficult to measure. Thus selection of animals to improve their fertility can be achieved by using BCS. It is also a marker for efficient maintenance of milk yield, milk quality and foetal development during lactation and gestation period. In beef breeds, body conformation of an individual relies on deposition of body fat and BCS acts as a primary indicator. As a whole, it helps in making decision of optimum management practices. Body condition score provides a biological relationship between body fat, milk production and reproduction. This relationship has economical implications with respect to dairy and beef cattle (Kadarmideen, 2004).

Phenotyping of BCS

Scoring is allocated with different scales by visual or tactile observations or both by different stakeholders like farmers, veterinarians, field workers etc. BCS can be recorded once or several times with manual observations using BCS chart and digital images over the lactation (Wildman et al., 1982). There are certain skeletal check points to estimate body condition score (Figure 1). These are: short ribs, hook bone, pin bone, Thurl, sacral ligament and tail head ligament. Scores are recorded separately observing each check points, then an average body condition score is assigned. There are several BCS charts available with various scales and most commonly adopted chart for dairy and beef cattle are depicted in table 1 and table 2; respectively. Phenotyping of BCS is performed on a 1- 5 scales in dairy cattle and 1- 9 on beef cattle. Ferguson *et al.* (2006) observed that assessment of BCS in dairy cattle is possible from digital photographs or video taken from rear of a cow rather than anterior and lateral view at 0-20° angle relative to tail head. Bell *et al.* (2018) also supported digital BCS and concluded that it provides accurate BCS and remove operator error. Both 2D and 3D camera can be used and picture can be analysed by visual image analysis system. Several applications like BCS Cowdition and BCS tracker are available to measure BCS.

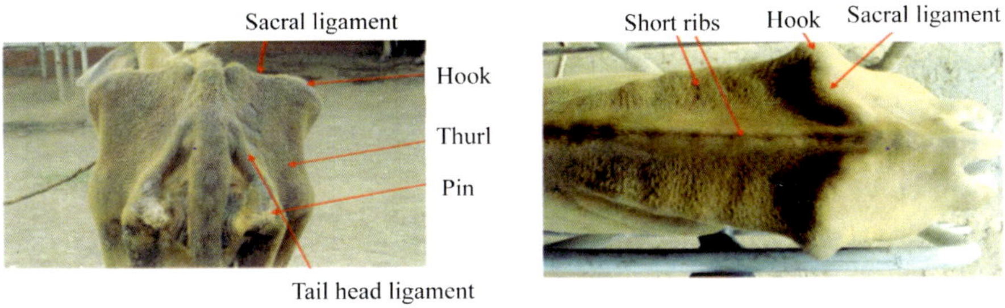

Fig. 1: Skeletal check points used to evaluate body condition (Gupta, 2008)

Ferguson *et al.* (2006) observed that assessment of BCS in dairy cattle is possible from digital photographs or video taken from rear of a cow rather than anterior and lateral view at 0-20° angle relative to tail head. Bell *et al.* (2018) also supported digital BCS and concluded that it provides accurate BCS and remove operator error. Both 2D and 3D camera can be used and picture can be analysed by visual image analysis system. Several applications like BCS Cowdition and BCS tracker are available to measure BCS.

Table 1: Body condition score chart for dairy cattle adopted from Ferguson et al. (1994)

BCS	Tuberosity-fatpad		Thurl	Ligament		Spinous process of short ribs visible
	Ileal (hook)	Ischeal (pin)		Coccygeal	Sacral	
<2.0	Angular-none	Angular-none	V	S	S	Angular >8 cm
2.25	Angular-none	Angular-none	V	S	S	Angular >8 cm
2.5	Angular-none	Angular-fat pad	V	S	S	Angular 6-8 cm
2.75	Angular-fat pad	Rounded-fat pad	V	S	S	Angular 6-8 cm
3.0	Rounded-fat pad	Rounded-fat pad	V	S	S	Angular 6-8 cm
3.25	Rounded-fat pad	Rounded-fat pad	U	S	S	Angular 4-6 cm
3.5	Rounded-fat pad	Rounded-fat pad	U	B	S	Angular 4-6 cm
3.75	Rounded-fat pad	Rounded-fat pad	U	B	B	Rounded 0-2 cm
4.0	Rounded-fat pad	Rounded-fat pad	U	NV	NV	Rounded 0-2 cm
>4.0	Rounded-fat pad	Rounded-fat pad	U	NV	NV	Rounded 0 cm

S-sharp, B- Barely visible, NV-Not visible

Table 2: Body condition score chart for beef cattle adopted from Richards et al. (1986)

BCS	Condition	Projection of tail-head and ribs	Spinous process with fat cover
1	Emaciated	Quite prominent	No fat and very sharp
2	Poor	Less prominent	Some fat and sharp
3	Thin	Identifiable, but not quite sharp	Obvious fat and less sharp
4	Borderline	No longer visible	Fat cover and Rounded
5	Moderate	No longer visible	Fat cover and spongy
6	High moderate	No longer visible	High fat cover and firm pressure to touch
7	Good	No longer visible	Very high fat cover and very spongy
8	Fat	No longer visible	Large fat deposits and impossible to palpate
9	Extremely fat	No longer visible	Extreme fat deposits and impossible to palpate

Genetic Analysis of BCS

Genetic and phenotypic parameters of BCS need to be estimated to assess its economic importance. h^2 of BCS and correlation estimates with other traits show genetic merit of an animal or of a herd. Effects of non-genetic factors are estimated as fixed effects and effects of genetic factors are considered as random effects. There are several approaches for estimating genetic parameters of BCS *i.e.*, h^2, genetic and phenotypic correlations with other economic traits. Those are: REML, AIREML, DFREML, MTDFREML, Animal, Sire, Mixed, Random Regression, Least Square, Bayesian etc. Genetic analysis for daily BCS, BCS at different stages such as BCS at service and calving, BCS loss can be estimated. Correlation between BCS and various economic traits indicates synergistic or antagonistic effects between the traits (Bastin and Gengler, 2013), presented in Table 3. h^2 can be estimated using univariate model, whereas, correlation can be estimated using bivariate model and then breeding value is estimated. Body energy change is environmentally and genetically driven and it is a moderately heritable trait. Its heritability is the least at the beginning of lactation and highest in mild to late lactation (Dechow et al., 2002). It appraises both the body composition and energy balance and is vital for metabolic stability, health and fertility (Pryce et al., 2001).

Body condition scores in first 3 lactations are highly correlated. Hence, only first lactation BCS can be used for easier approach. BCS is considered as an intermediate optimum trait to optimize milk production and the mean while minimizing metabolic and reproductive diseases (Bewley et al., 2008). Therefore, the frequency of genotypes with optimal conditions should be increased rather than over fat or underweight genotypes to fulfil satisfactory levels of yield and not over utilizing tissue reserves (GaliÇ, 2017). Bastin and Gengler (2013) concluded $h^{2\ of}$ BCS in the range of 0.14 to 0.51 with moderate inheritance.

Genotyping of a Population and Association of Genotype with BCS

Target population can be genotyped using various platforms of SNP array or sequenced with next generation sequencing methods. After obtaining common markers (SNPs), quality control is performed to collect good and informative SNPs and further association analysis is performed.

Table 3: Correlation estimates between BCS and various traits

Traits	Association with BCS
Production	
Milk yield, Fat yield, Protein yield, SCS	Negative
Fat%, Protein%	Positive
Type	
Dairy form, dairy character, angularity	Negative
Strength, Stature, Heart girth, Body depth	Positive
Body weight	Positive
Health and disease	
Mastitis	Negative
Diseases other than mastitis	Negative
Fertility	
Days to first heat, first service, Days open, Calving Interval	Negative
Conception rate	Positive

Association study explores genomic regions associated with phenotype variations in a target population to realize the genetic architecture of the phenotype (Mancin et al., 2022). Genome wide association study (GWAS) analyses were performed for BCS using univariate linear mixed model as used for genetic parameter estimation and Bonferroni correction for multiple testing. Individual significant SNPs in the GWAS were further analysed in a single marker association analysis where the SNP genotype was fitted as a fixed effect into the same univariate mixed model. So far only few genome-wide association study (GWAS) of BCS traits among livestock species have been carried out (Table 4).

Univariate Linear Mixed Model

$y = X\beta + Za + e$

where, y = trait EBV;

X = Vector of genotype;

β = Effect of SNP on trait EBV;

a = vector of random additive genetic effect;

Z = Incidence matrix;

e = random error

Table 4: Association study of BCS traits in livestock

Species	Animal (n)	Association	Reference
Sheep	Roman ewes (1034)	48,513SNPs QTL associated with body reserve (BR) present on chr 1 LEPR gene for adaptive capacity	Macé et al., 2021
Poultry	Chicken ecotype: Horro (384) & Jerso (376)	SNPs on chr 8 for BCS (Horro) *LMO4* (chr 8) & *LIMK2* (chr 15) genes	Psifidi et al., 2016
Cattle	Holstein	DGAT1polymorphism associated with milk production traits, BCS & its changes KK genotype (DGAT1 K232A polymorphism) shows higher fat content in milk, but no significant effect on BCS change	Kadlecova et al., 2014

SNPs and candidate region Annotation

All significant SNPs identified in the GWAS should be mapped to the reference genome of the respective species. Genes located in the vicinity of significant SNPs have to be identified within a window size based on the LD structure. Thereafter, the list of the identified genes is prepared. This gene list is further utilized for pathway analysis and Gene ontology (GO) terms enrichment for the structural and functional annotation. QTLs linked to these SNPs can be identified using Animal QTL database (Hu et al., 2013). Work-flow for genetic analysis of Body Condition Score and its genome wide association is presented in Figure 2.

Collection of data of BCS and other economic traits

↓

Analysis of effect of various genetic and non genetic factors

↓

Blood sample collection from different BCS groups

↓

DNA isolation, quantity and quality checking

↓

SNP genotyping

↓

lity control of genotype data

↓

Association study using univariate linear mixed model

↓

Annotation of associated significant SNPs

Fig. 2: Work-flow for genetic analysis of BCS and its GWAS

Conclusion

Body reserve levels can be assessed through body condition score (BCS) and body reserve dynamics through BCS changes over time between physiological stages. Body condition of cows are important for normal body growth and related with their production and reproduction efficiency. It also indicates stayability of the animal in a breeding herd and acts as a management technique used routinely to appraise the body fat reserves and energy status. Optimum BCS group for various economic traits can be identified. Thus, BCS can be considered in selection index with appropriate weightage and used for selection of animal in a herd for production and reproduction traits. Genome wide association study (GWAS) will be helpful for the identification of the candidate genes and QTLs linked to BCS. Further, SNPs for BCS can be included for genomic evaluation and genomic selection.

References

Bastin, C., & Gengler, N. (2013). Genetics of body condition score as an indicator of dairy cattle fertility: A review. Biotechnol. Agron. Soc. Environ., 17, 64–75.

Bell, M. J., Maak, M., Sorley, M., & Proud, R. (2018). Comparison of Methods for Monitoring the Body Condition of Dairy Cows. Frontiers in Sustainable Food Systems, 2.

Bewley, J. M., Peacock, A. M., Lewis, O., Boyce, R. E., Roberts, D. J., Coffey, M. P., Kenyon, S. J., & Schutz, M. M. (2008). Potential for Estimation of Body Condition Scores in Dairy Cattle from Digital Images. Journal of Dairy Science, 91(9), 3439–3453.

Dechow, C. D., Rogers, G. W., & Clay, J. S. (2002). Heritability and Correlations Among Body Condition Score Loss, Body Condition Score, Production and Reproductive Performance. Journal of Dairy Science, 85(11), 3062–3070.

Edmonson, A. J., Lean, I. J., Weaver, L. D., Farver, T., & Webster, G. (1989). A Body Condition Scoring Chart for Holstein Dairy Cows. Journal of Dairy Science, 72(1), 68–78.

Ferguson, J. D., Azzaro, G., & Licitra, G. (2006). Body condition assessment using digital images. Journal of Dairy Science, 89(10), 3833–3841.

Ferguson, J. D., Galligan, D. T., & Thomsen, N. (1994). Principal Descriptors of Body Condition Score in Holstein Cows. Journal of Dairy Science, 77(9), 2695–2703.

GaliÇ, A. (2017). Determination the Body Condition and the Relationship with Milk Yield of Turkish Holstein Cows. Tarım Bilimleri Dergisi, 23(4).

Gupta, P. (2018). Studies in the Body Condition Score in pure bred Sahiwal cows. M.V.Sc. thesis submitted to Indira Gandhi Krishi Vishwavidyalaya, Raipur.

Hu, Z.-L., Park, C. A., Wu, X.-L., & Reecy, J. M. (2013). Animal QTLdb: An improved database tool for livestock animal QTL/association data dissemination in the post-genome era. Nucleic Acids Research, 41(D1), D871–D879.

Kadarmideen, H. N. (2004). Genetic correlations among body condition score, somatic cell score, milk production, fertility and conformation traits in dairy cows. Animal Science, 79(2), 191–201.

Kadlecova, V., Němečková, D., Jecminkova Jochova, K., & Stádník, L. (2014). The effects of polymorphism in the DGAT1 gene on energy balance and milk production traits in primiparous holstein cows during the first six months of lactation. Bulgarian Journal of Agricultural Science, 20, 219–225.

Macé, T., González García, E., Foulquié, D., Carrière, F., Pradel, J., Durand, C., Douls, S., Allain, C., Parisot, S., & Hazard, D. (2022). Genome-wide analyses reveal a strong association between LEPR gene variants and body fat reserves in ewes. BMC Genomics, 23(1), 412.

Mancin, E., Tuliozi, B., Pegolo, S., Sartori, C., & Mantovani, R. (2022). Genome Wide Association Study of Beef Traits in Local Alpine Breed Reveals the Diversity of the Pathways Involved and the Role of Time Stratification. Frontiers in Genetics, 12.

Pryce, J. E., Coffey, M. P., & Simm, G. (2001). The Relationship Between Body Condition Score and Reproductive Performance. Journal of Dairy Science, 84(6), 1508–1515.

Psifidi, A., Banos, G., Matika, O., Desta, T. T., Bettridge, J., Hume, D. A., Dessie, T., Christley, R., Wigley, P., Hanotte, O., & Kaiser, P. (2016). Genome-wide association studies of immune, disease and production traits in indigenous chicken ecotypes. Genetics Selection Evolution, 48(1), 74.

Richards, M. W., Spitzer, J. C., & Warner, M. B. (1986). Effect of Varying Levels of Postpartum Nutrition and Body Condition at Calving on Subsequent Reproductive Performance in Beef Cattle23. Journal of Animal Science, 62(2), 300–306.

Siachos, N., Oikonomou, G., Panousis, N., Banos, G., Arsenos, G., & Valergakis, G. (2021). Association of Body Condition Score with Ultrasound Measurements of Backfat and Longissimus Dorsi Muscle Thickness in Periparturient Holstein Cows. Animals, 11.

Wildman, E. E., Jones, G. M., Wagner, P. E., Boman, R. L., Troutt, H. F., & Lesch, T. N. (1982). A Dairy Cow Body Condition Scoring System and Its Relationship to Selected Production Characteristics. Journal of Dairy Science, 65(3), 495–501.

Wright, I. A., & Russel, A. J. F. (1984). Partition of fat, body composition and body condition score in mature cows. Animal Production (UK).

14

Integrating Genomic Data into Breeding Programs for Cattle Improvement

T.V. Raja, Rani Alex, Umesh Singh, A.K. Das and Sushil Kumar

ICAR-Central Institute for Research on Cattle, Grass Farm Road
Meerut, Uttar Pradesh

India is bestowed with a wide variety of livestock species which forms the backbone of the economy of majority of the farmers. Cattle are one of the important species domesticated by man mainly for their draftability in agricultural operations. Rearing of cattle is an integral part of the Indian agriculture as it is always considered as a subsidiary enterprise of the agriculture farming. The indigenous cattle are well known for their adaptability to the harsh tropical climatic conditions, ability to thrive in the resource poor production systems, better feed efficiency and disease resistance. Barring few dairy breeds, majority of the Indigenous cattle breeds are poor in their genetic potential for milk production. Considering the ever increasing demand from milk , the country has implemented many cattle improvement programmes like Central Herd Registration Scheme (CHRS), Intensive Cattle Development Programme (ICDP), All India Coordinated Research Project (AICRP) on Cattle National Programme for Bovine Breeding and Dairy Development (NPBBD), National Livestock Mission, National Project for Cattle and Buffalo Breeding (NPCBB), Rashtriya Gokul Mission (RGM) etc., for improving the milk productivity of cattle.

Genetic improvement is one of the powerful tools for sustainable cattle production as the results obtained are cumulative and permanent (Mueller and Van Eenennaam, 2022). Unlike the nutritional, health and managemental interventions which needs continuous interventions, the genetic improvement made in one generation is being transmitted to the next generation. Thus, in country like India, implementation of the breed improvement programmes is essential to bring permanent increase in the milk production.

The success of any genetic improvement programme depends on a well-defined breeding programme with a definite breeding objective, accurate pedigree and performance recording for the identification and selection of genetically superior animals for the traits concerned. Traditionally, bull selection is considered as more important than the female selection as it contributes more number of progenies. Selection of superior bulls with better accuracy will help to bring the genetic improvement at the earliest time possible. The accuracy of selecting the elite bulls directly reflects the genetic gain and hence greater the accuracy of selection, the higher will be the genetic gain. The sires can be selected on the basis of the performance of their ancestors through pedigree selection or based on the performance of their daughters through progeny testing. The progeny testing is considered as a more reliable basis of bull selection. However, it is cumbersome, costly, and requires longer generation interval.

In India, cattle are raised mainly under small holder farming system with an average of two to five cows and well organized farms with large number of cattle are very scanty. This makes the progeny testing difficult under organized farm condition and hence the concept of associated herd progeny testing was developed through which more number of herds were included from different areas. Later the field progeny testing programme was started which envisages the testing of bulls under field conditions involving the farmers' herds was also developed to get the required number of daughters per bull for increasing the accuracy of selection.

In general, the bulls are evaluated under farm and field conditions to predict their genetic merit by estimating the expected breeding value (EBV). In spite of all these efforts, the accuracy of bull selection has not improved much to an extent comparable with the developed countries. The comparatively long generation interval and low fecundity are also some of the reasons for the slow genetic progress in cattle. Since the traditional selection programmes could not bring the required improvement at a faster rate, inclusion of information on the genetic makeup of the animals in sire selection would improve the accuracy of selection to a large extent. The genomic selection, a process by which the differences in the DNA information of the individuals are used to predict their genetic merit for selection will help to bring the desired genetic improvement with higher accuracy in a shorter period of time.

The advent of DNA sequencing and high-throughput genomic technologies together with the automated SNP genotyping resulted in a paradigm shift of selection strategies as the criteria moved from single gene/QTL to genomes, which can explain the majority of genetic variation in important traits. The inclusion of genomic data along with the pedigree information would improve the accuracy and the rate of genetic gain to a larger extent resulting in faster improvement of the trait concerned in the population. According to Culver and Labow (2002) genomics includes the process of sequencing, assembly and analysis of the structure and function of the entire genome. The advancement made in the field of genome sequencing and bioinformatics has resulted in better understanding of the biological phenomenon which is even the most complex (Kadakkuzha and Puthanveettil, 2013).

The molecular genetics research started with the candidate gene identification, followed by gene mapping, identification of QTLs and MAS, GWAS, genomic selection and the whole genome sequencing of the major livestock species (Rothschild and Plastow, 2014). The applications of individual gene identification, mapping of genes and identification of QTLs in cattle species have paved way for identification of various DNA markers to use them as selection criteria in the breeding programmes.

QTL Mapping and Marker Assisted Selection (MAS)

Quantitative trait loci (QTL) is a gene or a specific region of a chromosome which regulates or affects the quantitative trait or the loci which regulates the quantitative trait. The QTL will have an effect in a population only if it is in polymorphic state or have allelic variations and it must also be linked to a polymorphic allele which could be detected. The QTL mapping is the process of identifying the genomic regions or the markers which are significantly associated with the quantitative trait of interest. The implementation of marker assisted selection essentially involves two steps: 1. Identification of marker loci linked to QTL of economic importance and 2. Exploitation of linkage association in genetic improvement programmes. Once linkage between a marker locus and the QTL is established, by utilizing the marker alleles it is possible to recognize the alternative QTL allele inherited by the individual. Such information then can be used for the selection of breeding stock. MAS is likely to complement rather than replace the conventional breeding system leading to increased rate of genetic gain through increased selection intensity and reduced generation interval. Furthermore, selection on the basis of markers is possible in early life or in individual of both sexes for sex limited traits. However, there is a risk of reduced genetic response if this marker association information is inaccurate since MAS is a form of indirect selection.

Molecular markers unravel genetic variations either in the coding genes or in the non-coding sequence regions. On the basis of this characteristic the identification of markers influencing QTL can be carried out by two ways:

1. **Polymorphisms in the coding sequences:** DNA polymorphisms that occur in and around the structural or regulatory sequences of a genes of significance like milk protein genes, hormone genes, MHC etc., may directly affect the gene expression through changing the mRNA, rate of gene transcription or the sequence of gene product. So, therefore, it regulate phenotypic variations among the individuals in terms of productivity and health.
2. **Polymorphisms in the non-coding sequences:** In this approach, the variations occurring in non-coding regions like introns, flanking regions or in the intergenic regions of uncharacterized function) are utilized indirectly as markers for linkage analysis. These markers are often multiallelic and multiloci and so have high PIC. The association between the markers and the QTL is a function of distance between the markers and the genes controlling target traits, type of linkage phase and degree of linkage disequilibrium; high density gene maps with closer linkage are prerequisites for successful implementation of MAS (Smith and Smith, 1993).

Candidate Gene Approach

The candidate genes are defined as the genes with well-known biological function to regulate the development processes of the trait of interest directly or indirectly, which could be identified through association analysis. The basic assumption is that the individual variations in the phenotype of a quantitative trait are caused by the functional mutation of the putative gene regulating the trait. Candidate gene approach has been used widely for disease identification, drug targeting, genetic association studies etc., in many of the organisms. The success of candidate gene approach depends on the identification of casual relationship between the trait of interest and the candidate gene which is often difficult as this warrants the prior knowledge on the physiological, biochemical or functional mechanisms which are often very limited or sometimes may not at all available (Zhu and Zhao, 2007).

Genome Wide Association Studies (GWAS)

GWAS involves the identification of genomic variants which have statistically significant association with the trait of interest. In other words, it analyses the ancestrally related individuals with different phenotypes caused by the differences in the alleles or genotypes controlling it to establish the association between the phenotypes and genotypes (Uffelmann et al., 2021). The process of GWAS involves the recording of phenotypic information of individuals and the collection of blood samples for DNA isolation followed by genotyping either by using sequencing strategies or the GWAS arrays, imputation of the untyped variants using the reference population, association study using appropriate statistical test for interpretation of the results.

Genomic Selection

Genomic selection or whole-genome selection is proposed by Meuwissen and co-workers in 2001, as they demonstrated the possibility to make very accurate selection decisions when breeding values were predicted from dense marker data alone. Genomic selection is a sort of marker-assisted selection wherein genetic markers covering the entire genome are employed in a bid to ensure that all QTL are in linkage disequilibrium with at least one marker. This ensures that the most accurate results may be obtained (Goddard and Hayes, 2007). The genomic selection helps to associate the effect of variations in a large number of SNPs (in thousands) for different production traits. As and when the DNA profile of an animal or young bull is available, the GEBV is predicted as the sum of all DNA variants (SNPs) it possesses which indicates the genetic merit of that individual.

Implementation of genomic selection can be done in three stages discovery, validation and selection. To derive and validate the prediction equation based on the SNP is the preliminary stage of genomic selection. The discovery and validation stages are employed in this regard. In the discovery or training generation a large number of SNP are assayed on a moderate number of animals having phenotypic records for all relevant traits which help to generate discovery data set and based on this a prediction equation is developed, that uses the markers as input and predicts the breeding value derived from this data. In the validation set, genotyping is carried out at least for the markers that are to be used for commercial purpose, but in a large number of animals whose phenotypic information is available. The accuracy of prediction equation is also tested on the independent sample in the stage of validation. The prediction equation developed will be used to estimate the genomic estimated breeding value (GEBV) of individuals for which the individuals will be genotyped for different markers.

Genomic selection (GS) can increase the response to selection in three different ways, by increasing the accuracy and intensity of selection as well as by decreasing the generation interval. The accuracy of selection by genomic selection is further dependent on some other factors such as the level of LD between markers and the QTL, the number of animals in the reference population, the heritability of the trait in question and the distribution of QTL effects. A simulation investigation that involved financial implications showed that dairy cattle breeding organisations may save up to 92% of breeding expenditures if the conventional progeny test system was substituted with a GS breeding programme. This was shown in the study (Schaeffer, 2006). The drastic decrease in the generation interval and the rise in the selection accuracy for bull dams are primarily responsible for this significant cost savings.

More than 16 countries have included GS into their dairy cattle breeding programmes, both on a national and international scale (Nilforooshan et al., 2010; Eggen, 2012). But in other species like beef cattle, sheep and goat, the implementation is still in preliminary stages. Further, in the developed countries, the whole genome sequence analysis and genomic selection have already been used in the selection schemes of the breed improvement programmes. However, in countries like India, the use of genomic data in the breeding programme is in its primitive stage.

Reference Population

Different government and non-government organizations who are presently engaged in progeny testing programme will be included in the project. Since it is very much challenging to create a reference population from a single breed or population, population from different breeds will be combined to develop a reference population for multi-breed genomic selection (which includes Sahiwal, Gir, Kankrej, Tharparakar, Ongole and other indigenous breeds as well as Holstein crossbreds developed in India over a period of time). The biological tissues either cryo-preserved semen or blood samples will be collected from different organizations. The bulls evaluated over a period of time will be used as reference population. The bulls will be classified into training and validation populations. The youngest 20% bulls will be included in the validation population and 80% of the oldest bulls will be included in the training population. Also selected cows from different populations also will be included in the validation set to increase the size of reference population. Further in order to increase the accuracy by inclusion of dominance effects the grand-daughter design will be followed.

Validation of Available HD and LD SNP Chips in Indian Populations

Since the project involves different breeds, identification of conserved LD across breeds to maintain the association between SNP and QTL is difficult with LD chips. So HD chips will be utilized for the screening training set of reference population. Further different commercial and customized LD chips with varying densities (from 3K and 50K) and SNP selection criteria will be tested. The best suited LD chip will be selected for further genotyping and will be imputed to HD. Imputation will be carried out using software Beagle 3.3.0 (Browning and Browning, 2007).

The WGS or high-density genotyping of all the individuals of each generation may not be economical and hence the SNP chips covering the SNPs of indigenous breeds can be used. So an attempt will be made to develop customized low density chip based on the identified SNPs and will be used in further genotyping in subsequent generations.

Development and Validation of Prediction Equation for GEBV

The prediction equation will be derived from the SNPs by dividing the entire genome into segments and the effects of which will be estimated in a reference population in which animals are both phenotyped and genotyped. Two methods will be used to derive the prediction equations viz., BLUP approach (Meuwissen et al., 2001) and Bayesian approach. Further the accuracy of the genomic selection will be validated in the validation population (20% of youngest bull). Here the predicted GEBV estimated from SNP effects will be correlated with the current breeding values of the bulls, which were largely derived from a progeny test. Using selection index theory, the final GEBV will be computed by combining the parental average breeding value from pedigree information with the breeding value from genomic information. The parental average or polygenic breeding value will capture the effects of QTL which are not captured by SNP effects.

Implementation of Genomic Selection in the Organized and Farmers' Herds

A hybrid breeding scheme with a conventional progeny testing breeding program and genomic selection will be most suitable in Indian conditions as explained by Thomasen et al. (2014) in Danish Jersey breeds. But due to the inclusion of different well defined indigenous breeds in India, the programme will be run separately for indigenous breeds and crossbreds. For this a component of genomic selection will be added to the existing progeny testing programmes in India. The implementation of genomic selection will be carried out differently in indigenous breeds and crossbreds. In the initial phase, genomic selection will be incorporated as a part of progeny testing programme in the breeds where herd strength is more than 10000 where proper recording is going on. In the case of indigenous breeds (Fig1A) a minimum of 500 cows will be declared as elite herd for the production of male calves.

These cows will be inseminated with relevant bull sires to produce 100 male calves. These male calves will be genotyped and 15 young bulls will be selected according to their GEBV. From these 5 young bulls with the highest GEBV will be selected as bull sires and will be mated to 30% of the bull dams. Remaining 10 bulls will be randomly used in 75 percent of population for progeny testing. Finally 2 proven bulls will be selected both for use in bull dams, contributing 70% of the inseminations, and for inseminations in the cow population, and contributing 25% of the population.The implementation of genomic selection in the crossbreds will be in a similar way but with higher selection intensity which is demonstrated in Figure 1B.

Necessity of Genomic Selection in India

In India, most of livestock are reared under small holder production systems and are always limited by small herd, challenging environment, low inputs and poor management and uncontrolled mating. The increase of indiscriminate cross breeding is another challenge to the animal breeders. The alarming fact is that systematic production records and herd registers are not prepared for approximately 90% of cattle, buffaloes and is practically do not exist for other species or limited to few breeding farms. The systematic data recording for Livestock in India exists only in Institutional, Government, Dairy Co-Operatives and few Farms owned by govt. institutions and elite group of farmers. The fact remains that our country has one of the highest populations of cattle and breeds available within the species are also very large when compared to European and American continents.

Under such conditions, the genomic selection programme based on the genomic information of animals would be useful for early selection of animals with increased accuracy. The equation for GEBV prediction already been developed in the reference population can be very well used under field conditions for selection of animals without any phenomic data. It will also help to increase the accuracy of selection thereby increase the genetic gain.

Conclusion

The integration of genomic data into the breeding programmes help to improve the genetic gain in the trait of economic importance at a faster rate. It also helps to increase the accuracy of selection, reduce the generation interval and increase the intensity of selection for faster genetic gain. However, the cost of genomic data generation by WGS or NGS is the major concern for implementation under Indian conditions. Further, the availability of reliable phenotypic data is also one of the major concerns which limits the use of genomic data in the breed improvement programmes.

References

Browning S.R. and Browning B.L. (2007). Rapid and accurate haplotype phasing and missing-data inference for whole-genome association studies by use of localized haplotype clustering. American Journal of Human Genetics 81(5): 1084-97.

Culver, K. W. and Labow M.A. (2002). "Genomics," in Genetics, ed R. Robinson (New York, NY: Macmillan Science Library, Macmillan Reference USA).

Eggen, A. (2012). The development and application of genomic selection as a new breeding paradigm. Animal Frontiers 2: 10–15.

Goddard M. and Hayes B.J. (2007). "Genomic selection." Journal of Animal Breeding and Genetics 124 (6): 323-30.

Kadakkuzha B.M. and Puthanveettil S.V. (2013).Genomics and proteomics in solving brain complexity. Molecular Biosystems 9: 1807–1821.

Meuwissen T.H., Hayes B.J. and Goddard M E (2001).Prediction of total genetic value using genome-wide dense marker maps. Genetics 157: 1819–1829.

Mueller M.L. and Van Eenennaam A.L. 2022. Synergistic power of genomic selection, assisted reproductive technologies, and gene editing to drive genetic improvement of cattle. CABI Agriculture and Bioscience 3(1):13.

New York, NY: Macmillan Science Library, Macmillan Reference USA) (New York, NY: Macmillan Science Library, Macmillan Reference USA)

Nilforooshan M.A., Zumbach B., Jakobsen J., Loberg A., Jorjani H. and Durr J. (2010). Validation of national genomic evaluations.Interbull Bulletin 42: 56-61.

Rothschild M.F. and Plastow G.S. (2014).Applications of genomics to improve livestock in the developing world. Livestock Science 166: 76–83.

Schaeffer L.R. (2006). Strategy for applying genome-wide selection in dairy cattle. Journal of Animal Breeding and Genetics 123: 218-223.

Smith C. and Smith D.B. (1993). The need for close linkages in marker assisted selection for economic merit in livestock. Animal Breeding Abstract 61:197-204.

Thomasen J.R., Egger-Danner C., Willam A., Guldbrandtsen B., Lund M.S. and Sorensen A.C. (2014). Genomic selection strategies in a small dairy cattle population evaluated for genetic gain and profit. Journal of Dairy Science 97 (1): 458-470.

Uffelmann, E., Huang, Q.Q., Munung, N.S. et al. Genome-wide association studies. Nat Rev Methods Primers 1, 59 (2021).https://doi.org/10.1038/s43586-021-00056-9.

Zhu M. and Zhao S. (2007). Candidate Gene Identification Approach: Progress and Challenges. International Journal of Biological Sciences, 3 (7): 420-427.

15

Enhancing Genetic Gain Through Powerful Techniques Genomic Selection, Assisted Reproductive Technologies and Gene Editing

***Rajesh Kumar Vandre*[1], *Amit Kumar Jha*[1], *Sulochana Sen*[1] *and Bhabesh Chandra Das*[2]**

[1]*Department of Animal Genetics and Breeding, College of Veterinary Science & Animal Husbandry, Rewa, Madhya Pradesh*
[2]*Department of Veterinary and Animal Husbandry Extension Education College of Veterinary Science & Animal Husbandry, Rewa, Madhya Pradesh*

Introduction

The key factor of sustainability in the animal sector depends upon genetic improvement in domestic livestock. There are many reproductive and molecular biotechnologies used for the genetic improvement of domestic livestock and they are synergistically coupled with the breeding program in distinct breeding targets. With the help of gene editing techniques (e.g. CRISPR etc.). To improve the specific traits of interest scientists can add, delete or replace the nucleotides (e.g. milk production, meat production, disease resistance, etc.) in less than one generation. Gene editing with the conventional breeding program in livestock is an important factor for genetic improvement for increasing the rate of genetic gain. In the present topic, here, we delved into the most important reproductive and molecular technologies used in livestock genetic improvement, as well as the most effective ways for incorporating gene editing into existing genetic improvement programmes and targeting specific traits for enhancement. The improvement in genetic level is a most powerful part of sustainable animal development because it is permanent and the effects are long-lasting and cumulative. Genetic advancements produced in one generation are passed on to the following, unlike dietary and animal health treatments, which are a continuous process. The many livestock operations that involve chemical and mechanical methods such as dehorning requires time, worker, money, animal safety, etc., but their genetic alternation of the concerned gene for the polled or hornless condition can exclude the requirement of physical dehorning. (Gottardo *et al.* 2011; Tompson *et al.* 2017).

Production efficiency and sustainable development in the animal sector are increased with each other. Efficiency may be defined as constructing more with even fewer resources. Livestock improvement started with selective breeding with estimated breeding value (Statistical Prediction) now in the present era genomic selection blended with assisted reproductive technologies has able to more accurate selection and utilization of highly productive parents for the production of the next generation and improvement in average genetic value in a population.The genetic improvement programsfor domestic livestock begin with traditional breeding, assisted reproductive technologies (ART), and genome editing. With the help of this only a small number of animals, genetic improvement makes high production possible without a vast population. The livestock population's impact on the environment can also minimize with genome editing. The effect of increasing rates of genetic gain is improvingsustainable livestock production. The study on artificial insemination over natural service estimated that about half of the increased productivity per animal increased the genetic gain rate (Bertolini and Bertolini, 2009). The breeders are used the equation to measure the genetic gain in livestock breeding programs. The equation of rate of genetic gain is (Lush, 1937)

$$(\Delta G) = h^2S = (i \times r \times \sigma A)/ L$$

Where

h^2 = heritability

S = Selection differential

i = Selection intensity

r = Selection accuracy

σA = Genetic diversity

L = Generation interval

If the numerator's components are increased and the denominator, or generation interval, is decreased, the pace of genetic gain may be accelerated.With the aid of genomic information, the precision of selection is improved in comparison to pedigree information alone. Reproductive and molecular biotechnologies, like assisted reproductive techniques and genomic selection, can be used in a systematic breeding program to speed up rates of genetic gain.

Artificial insemination (AI) and embryo transfer technology (ET) are the techniques that have been included in livestock breeding programs to enhance selection intensity and are known as assisted reproductive techniques. In parallel, the development of advanced genotyping of single nucleotide polymorphism (SNPs) has made it possible for genomic selection (GS) to determine an animal's genetic worth based on its genetic contents (DNA) (Meuwissen *et al.* 2001).The generation interval has been shortened while selection precision has increased because of the use of GS. Additionally, previous studies reported that GS can provide information on qualities that are difficult or expensive to record, are reported late in life, or both (Garcia-Ruiz *et al.,* 2016; Hayes *et al.,* 2013 and Meuwissen *et al.,* 2013). One of the most recently created technologies for genetic enhancement is genome or gene editing is CRISPR-cas9. Breeders are now able to very accurately pinpoint the insertion, deletion, or substitution of base pairs in the genetic code in order to impact desired traits. This is made possible with the assistance of cutting-edge biotechnology.

Genomic Selection

There is a major impact of genomic selection in animal breeding programs. It easily very much helpful to the traits which are difficult to select in many situations. Genomic selection increases selection accuracy and enables shorter generation intervals by providing more precise estimates of breeding value early in the life of breeding animals. From small to large ruminant animals, the genetic improvement rate may vary and it is very tough to measure the traits which are improved.

The rate of gain through genomic selection may be double. In some species such as buffalo, pigs, goats and sheep, growth trait is achief trait and this should be measured before reproduction. Hard to assess traits like meat quality, feed effectiveness, reproduction, and disease resistance can all be increased more effectively through genomic selection. There is a requirement of a reference population to predict breeding values based on genomic information (thousands of animals measured) and the efficient use of genomic selection can be possible with the required size and genetic makeup of a reference population.

The accuracy in genomic selection depends upon the size of the population, effective population size, and heritability of the animal. The number of sires utilized in breeding populations is mostly related to the effective population size (Ne), which is used to identify how much common genes shared by the not related individuals within a population. If the effective population is low, it means greater the shared DNA segments. The previous study of actual data and simulation modeling both showed that the accuracy of genomic predictions decreases as the distance between the predicted animals and the reference population increases. A pedigree-based BLUP prediction of breeding value will be fairly accurate if an animal has a large number of closely related animals in the reference population. The predictions will depend on "average relatedness" between the individuals of population if they have no direct relatives in the reference group. (Goddard *et al.,* 2009).

There are many technologies and new statistical toolse.g. whole genome sequencing, genome-wide signatures of selection etc are the dominant tool for selection and obtaining high genetic gain. The whole genome sequence analysis and genomic selection (GS) are used in breeding schemes of domestic animals (cattle, sheep, goats, chicken, pigs) in developed countries butother than in developing countries, technologies such as genome study are applied for assessing genetic variability, polymorphism study, selection to identify genomic regions etc.

Teamwork is required between developing and developed countries to implementthe genomic breeding technologies for getting high genetic gain and response. Studied showed that there were some possible impacts happened in collaboration works Li *et al.* (2015).

Assisted Reproductive Technology (ART)

There are some technologies known as assisted reproductive technologies that can increase the rate of genetic improvement but they may also increase the rate of inbreeding. With the use of these technologies how to take advantage, while at the same time minimizing their genetic disadvantages is a major challenge for the breeder. Artificial insemination (AI), multiple ovulation and embryo transfer (MOET), and *in vitro* embryo production (IVEP) can increase the rate of genetic improvement, with tolerable rates of inbreeding. Semen sexing, embryo sexing, and embryo cloning, on the other hand, can only result in modest increases in the rate of genetic progress. Though with embryo cloning once onlysignificantly boost the genetic merit of marketable stock and may change breed structure. Reproductive technology is used in combination with the preservation of animal genetic resources. All animal species with very low populations can be preserved with the use of the cryopreservation technique.

The genetic effects of reproductive technologies increase reproductive potential. In comparison to natural service, we can argue that only a fewer parents are able to generate the necessary quantity of offspring. We can say that it increases the intensity of selection.

Artificial insemination is the technique in which highly proven siresareused to produce the required number of offspring. Artificial insemination can be used to increase sire selection intensity.Therefore it increases the average genetic gain of offspring. The tested sire semen can be sent to areas where natural service is not feasible by artificial insemination.

The process is embryos being collected from a donor female and transferred to a recipient female is known as embryo transfer technology. Embryo transfer technology is very much useful for the production of documented purebred animals or native animal populations. With the help of embryo transfer technology, genetically superior females can be used for the production of more offspring than natural production. Now a days embryo transfer technology is used for the conservation and production of several animals of domestic species such as cows, buffalo, goats, sheep, etc.The vital goal of animal raising is to increase each animal's productivity in a short amount of time at a low cost. Due to a number of difficulties, including a lack of high-quality genetic material in adequate amounts, a lack of breeding facilities, the small size of the herd, and issues with data collection and processing, the genetic rate of development is very low, making the use of embryo transfer technology useful for achieving our target goal. Traditional breeding methods for livestock production include the MOET-ONBS breeding scheme.

One another method known as semen sexing has to use for increasing the number of progeny of one sex in a population, thismay increase the intensity of selection for that particular sex. But this may cause reduce the other sex numbers, this can reduce the intensity of selection in the other sex.

If males and females are chosen based on their performance at a reasonably young age, the rate of genetic improvement is relatively low (McClintock and Nicholas, 1991; Bekman *et al.,* 1994; Villanueva *et al.,* 1995). The study reported that the use of sexed semen can increase the rate of genetic improvement by up to 25% in a traditional large-scale dairy cattle progeny-testing scheme (Van Vleck, 1981).

Gene Editing

The technology of altering genomes can speed up the process of breeding for genetic improvement (Jenko et al., 2015). Genome editing refers to the process in which specific sections of a cell's genome can have alleles added, deleted, or changed in order to achieve a desired result. When changes occur in somatic cells, they are temporary; but, when alterations occur in zygotes or germ-line cells, they are permanent and inherited.There are more than 300 examples of plants and animals that have had their genomes edited (Carroll *et al.,* 2015).

The favorable alleles can be fixed more quickly within breeding programs with genome editing technology. One method of genome editing is gene drives. The term "gene drives" refers to mutations that arise naturally on one chromosome and cause that chromosome to copy itself onto the homologous chromosome. Initiation of Double-stranded breaks occurs on the homologous chromosome via gene drive. The study on gene drive was reported in Drosophila melanogaster in 1950 on P-element (Clark *et al.,* 1997). To ensure that all offspring are homozygous for the modified cell, the gene drive can be used in combination with genome editing technology on either the parent's germ cells or the zygotic cell stage.

Targeted gene drive mechanisms conversion efficacies of more than 98 percent have been reported for the newest gene editing technique, CRISPR/Cas9 (Gantz*et al.,* 2015) demonstrating how this technology might be used to transmit edited genes throughout populations. Genome editing combined with gene drives for quantitative traits loci to fix edited genes more rapidly in livestock populations. Therefore, we can infer that modifying the animal genome can be utilized to significantly increase the frequency of beneficial genes. The gene drive can increase the altered genes throughout the livestock populations.

Enhancing large animal fitness and productivity, boosting disease resistance, promoting animal welfare by enhancing animals' resilience and adaptation, and controlling pests and ticks that are harmful to livestock could be achieved through genome editing/CRISPR cas9 technology.

Innovative solutions to troubles related to livestock populations, such as productivity and fitness traits, the health and well-being of animals, environmental preservation, and effects on human health, have been made possible by the introduction of CRISPR for genome editing.

With the use of this technology, livestock species and the environment can be changed in manner never before conceivable in order to achieve human goals and improve overall wellbeing.

Without a doubt, the development of gene editing technology has been of immense value to the development of agricultural and livestock species breeding. In any organism, a new variant can create by DNA base-level editing. Disease resistance gene/genes can be inserted or fixed with the help of gene editing technology that is not possible with traditional breeding techniques.

We can say that the advent of gene editing technology has produced immeasurable value for the progress of livestock breeding. The variation can be createdby DNA base-level editing into individuals or species.

The development of the CRISPR/Cas system represents yet another significant advance in the precise and effective editing of particular genes in the mammalian genome. CRISPR/Cas system uses nucleic acid to select targets but ZFN and TALEN, which employ protein to select targets. There are some constraints of CRISPR/Cas technology such as possible off-target effect, cytotoxicity, stringent PAM constraints, and low pathway efficiency, etc (Sternberg *et al.,* 2014; Fu *et al.,* 2013; Heyer *et al.,* 2010). The future of this technology will depend toconstrain on how well these issues can be resolved.

Opportunity

The application of genomics to breeding in livestock has had a major economic impact (i.e., genomic selection, GS). To acquire precise genomic breeding values, the sufficient number of animal diversity and population is required. The strategies for breeding combining with conventional and genomic data provide positive outcomes.

Conclusion

With the capability of technology like genome editing, assisted reproduction, and genomic selection, response to selection or genetic gain may greatly increase. In the current situation, the nation needs these kinds of technology in order to meet the need for food, for which humans depend on livestock. With these methods, it may be able to improve genetic gain while decreasing the generation gap. Researchers and scientists should investigate this topic in the current context in order to upgrade and improve the livestock and agricultural fields.

References

Bekman, H., Meuwissen, T.H.E. and Oldenbrcek, J.K., (1994). Increasing beef production from dairy cows by implanting embryos from MOET nucleus breeding schemes for beef cattle. Livestock Prod. Sci., 37: 271-282.

Bertolini, M. andBertolini, L (2009). Advances in reproductive technologies in cattle: from artifcial insemination to cloning. RevisFac Med Vet Zootecnia.;56(3):184–94.

Carroll, D. and Charo, R.A. (2015). The societal opportunities and challenges of genome editing. Genome Biol.,16:242.

Clark, J.B., Kidwell, M.G. (1997). A phylogenetic perspective on P transposable element evolution in Drosophila. ProcNatlAcadSci USA., 94:11428–33.

Fu, Y., Foden, J.A., Khayter, C., Maeder, M.L., Reyon, D.,Joung, J.K. and Sander, J.D. (2013). High-frequency off-target mutagenesis induced by CRISPR-Cas nucleases in human cells. Nat. Biotechnol., 31, 822–826.

Gantz, V.M., Jasinskiene, N., Tatarenkova, O., Fazekas, A., Macias, V.M., Bier, E.(2015). Highly efficient Cas9-mediated gene drive for population modification of the malaria vector mosquito Anopheles stephensi. ProcNatlAcadSci USA., 112:E6736–43.

García-Ruiz, A., Cole, J.B., VanRaden, P.M., Wiggans, G.R., Ruiz-López, F.J., and Van Tassell, C.P.(2016). Changes in genetic selection diferentials and generation intervals in US Holstein dairy cattle as a result of genomic selection. ProcNatlAcad Sci.;113(28):E3995–4004.

Goddard, M.E. (2009). Genomic selection: prediction of accuracy and maximization of long term response. Genetica, 136:245–257

Hayes, B.J., Lewin, H.A. and Goddard, M.E (2013). The future of livestock breeding: genomic selection for efciency, reduced emissions intensity, and adaptation. Trends Genet.;29(4):206–14.

Heyer, W.D., Ehmsen, K.T. and Liu, J. (2010). Regulation of homologous recombination in eukaryotes. Annu. Rev. Genet. 44, 113–139

Jenko, J., Gorjanc, G., Cleveland, M.A., Varshney, R.K., Whitelaw, C.B., Woolliams, J.A., (2015). Potential of promotion of alleles by genome editing to improve quantitative traits in livestock breeding programs. Genet SelEvol., 47:55

Lenstra, J. L., Groeneveld, L. F., Eding, H., Kantanen, J., Williams, J. L., Taberlet, P., et al. (2012). Molecular tools and analytical approaches for the characterization of farm animal genetic diversity. Anim. Genet. 43, 483–502. doi: 10.1111/j.1365-2052.2011.02309.x

Lund, Li, X., Zhang, M. S., Costa, Q., Ducrocq, C. N., and Su, G. (2015). Short communication: improving accuracy of predicting breeding values in Brazilian Holstein population by adding data from Nordic and French Holstein populations. J. Dairy Sci. 99, 4574–4579. doi: 10.3168/jds.2015-10609

Lush, J.L. (1937). Animal Breeding Plans. Ames, IA: Collegiate Press, Inc.;

McClintock, A.E. and Nicholas, F.W. (1991). The implications of advanced breeding techniques. Project No. USO16. Final Report, Australian Meat and Livestock Research and Development Corporation, Sydney.'

Meuwissen, T., Hayes, B. and Goddard, M. (2013). Accelerating improvement of livestock with genomic selection. Annu Rev AnimBiosci.1(1):221–37.

Meuwissen, T.H.E. and Hayes, B.J.(2001). Goddard ME. Prediction of total genetic value using genome-wide dense marker maps. Genetics.;157(4):1819–29

Sternberg, S.H., Redding, S., Jinek, M., Greene, E.C. and Doudna, J.A. (2014). DNA interrogation by the CRISPR RNA-guided endonuclease Cas9. Nature, 507, 62–67.

Thompson, N.M., Widmar, N.O., Schutz, M.M., Cole, J.B. and Wolf, C.A. (2017) Economic considerations of breeding for polled dairy cows versus dehorning in the United States. J Dairy Sci.;100(6):4941–52.

Utsunomiya, Y. T., Bomba, L. B., Lucente, G., Colli, L., Negrini, R., Lenstra, J. A., et al. (2014). Revisiting AFLP fingerprinting for an unbiased assessment of genetic structure and differentiation of taurine and zebu cattle. BMC Genet. 15:47. doi: 10.1186/1471-2156-15-47.

Van Vleck, L.D. (1981). Potential impact of artificial insemination, sex selection, embryo transfer, cloning, and selting in dairy cattle. In: B.G. Brackett, G.E. Seidel, Jr. and S.M. Seidel (Editors), New Technologies in Animal Breeding. Academic Press, New York, pp. 221-242.

Villanueva, B., Woolliams, J.A. and Simm, G., (1995). The effect of improved reproductive performance on genetic gain and inbreeding in MOET breeding schemes for beef cattle. Genet. Sel. Evol., 26: 347-363.

16

Genomic Selection in Different Livestock Species

Amit Baranwal[1] Avneesh Kumar[2] and Rajni Chaudhary[3]

[1]Krishi Vigyan Kendra, Pratapgarh, Uttar Pradesh
[2]Departtment of Animal Genetics and Breeding, DUVASU, Mathura, Uttar Pradsesh
[3]NTRS,GARSA, ICAR-CSWRI, Avikanagar, Rajasthan

The global human population is increasing exponentially with time, and it is expected to reach 9.6 billion by 2050 (UN 2013). With rapid population growth, food consumption is predicted to more than double in the next 40 years. Due to limited land resources for cultivation and a stabilising trend in agricultural productivity, the role of livestock sector is increasing significantly to meet the rising food demand. Meanwhile, global livestock production has increased significantly in both number and productivity. Unprecedented urbanisation has a significant impact on food consumption patterns in general, as well as demand for livestock products. Animal diseases are a significant factor in reducing livestock productivity, particularly in developed countries. The limitations of traditional animal breeding hinder further improvement of livestock productivity. The use of biotechnological tools has major applications in improving animal health and production, increasing disease resistance, using assisted reproductive techniques, improving rapid diagnostic techniques, and improving vaccines in livestock systems. These interventions cover advancements in feeding, nutrition, genetics, reproduction, animal health control, and overall improvement of animal husbandry. Since genomic tools and selection have been widely adopted in developed countries to improve both productivity and quality of animal products, the prospects for increasing productivity in developing countries using genomic tools and selection have not been adequately investigated.

Traditional Breeding of Livestock

Traditional animal breeding, also known as TAB, is a method that assigns points to animals based on the phenotypes or outward appearances that can be observed of them. This allows the breeding value of an animal to be determined. Since most features (such as meat quality and milk production) are difficult to evaluate before an animal reaches maturity, this approach is frequently inaccurate, inefficient, and time-consuming, which delays the verification of breeding results. In TAB, breeders establish a breeding goal based on desired production attributes specific to a particular breed. Using either quantitative or qualitative criteria, sires and dams are phenotypically evaluated for each breeding generation. The scores for each attribute provide a breeding value (BV), which is used to rank the animals according to their genetic merit. Breeding is limited to sires and dams with the greatest BVs, or most desirable physical traits. Their kids are then appraised in accordance with the breeding goal specified by the breeder, and the individuals who have the highest BV are used as breeding material for the subsequent generations of animals.

Constraints of Traditional Animal Breeding

In TAB, breeders establish a target for breeding depending on the characteristics of offspring they hope to see. Each generation of breeding animals is examined phenotypically, either quantitatively or qualitatively. Based on their accumulated BVs, the animals are ranked in order of their genetic superiority. Only the most visually appealing sires and dams are allowed to breed. The children are evaluated according to the breeding goal, with the highest BV offspring being used for further generations. The progeny of two high BV animals will not necessarily have a greater BV, even if the phenotyping is perfect. Complex inheritance of traits, number of traits to be assessed in a breeding goal, a decrease in selection efficiency for the difficult to measure traits, lower heritability of reproduction traits, attributes to be measured in large population, traits with high selection pressure and slower genetic progress are the key variables to limit this process.

Marker-Assisted Selection: Complementation Rather Replacement

The development of new genetic tools to aid in breeding decisions has been accelerated by new sequencing technology. These early sequencing attempts could only represent a small portion of animal genomes due to time and expense constraints. Because of advances in genotyping and gene expression technology, researchers have been able to identify DNA regions associated with peculiar traits known as quantitative trait loci (QTL). Many small-effect genes or a few large-effect genes may be involved in QTL. Linkage disequilibrium (LD) mapping was used to construct linkage maps to reveal the relative locations of each QTL on the genome (determining the correlation of these markers within a population). Further researchers discovered genetic markers and created marker panels for each QTL, which could be used in performing marker-assisted selection (MAS). Marker assisted selection MAS is an indirect selection process in which a trait of interest is chosen based on a marker associated with that trait rather than the trait itself. The first assumption for MAS is that the markers are associated at high frequency with genes or QTLs of interest due to genetic linkage. The second assumption is that the distance between markers and linked loci affecting traits should be as short as possible (within 5 cM) or within the gene of interest.

Methodology of MAS

↓

Identification of marker loci

↓

Establishment of linkage map of marker

↓

Detection of association between marker and QTL

↓

Exploitation of Linkage association in genetic improvement

Despite of difficult, costly, time-consuming and technically demanding process for detection of marker loci and establishment of associations between markers and loci, limited number of markers for economic traits and lack of total genetic maps, marker assisted selection has prospects for selection that cannot be improved by conventional breeding program. MAS improves the efficiency of selection for traits with few records (due to expensive recording), low heritability traits, sex limited traits, carcass traits, disease resistance traits and survivability. It shortens the generation interval because animals can be selected at a young age, improving genetic gain. Therefore, MAS is likely to complement rather than replace conventional breeding system leading to an increased rate of genetic improvement. MAS offers breeders a starting point for utilising genomics to accomplish breeding objectives. Applications of MAS include marker assisted introgression (for example, the fecundity gene in sheep), the elimination of harmful gene alleles (for example, the fish taint in brown-shelled chicken eggs or the halothane in pigs), and the selection of optimal circumstances based on certain marker information (such as the polled condition in cattle).

Genomics: A Paradigm Shift in Livestock Breeding

The development and implementation of new genomic tools specifically improved DNA sequencing methods, increased availability of high-throughput genotyping platforms and commencement of human genome project were triggering factors for emergence of this genomic era. Next-generation sequencing (NGS) technologies are widely used for detection of millions of Single Nucleotide Polymorphisms (SNPs) and also provide a means of assessing their variation. This information can be used to create subsets of highly informative SNPs for genome-wide or region-specific analyses, as well as to find mutations that control phenotypic variation within or between populations. SNP identification through NGS ledto the development of Genome-wide panels of SNPs. The genetic merit of animals can be estimated by genomic selection, which uses genome-wide SNP panels as markers and statistical methods that capture the effects of large numbers of SNPs simultaneously. From a scientific perspective, rapid genome-scale measurements have a significant impact on our understanding of the fundamental structure and function of livestock genomes, as well as how complex characteristic characteristics are controlled. It accelerated This narrowed the discrepancy between qualities that are inherited and those that are observed and aid in bridging the genotype/phenotype gap. Comparison of whole genome sequences of several individuals of different breeds with a reference sequence produced a never-ending source of genetic markers, primarily single nucleotide polymorphisms (SNPs). Therefore, the availability of whole-genome SNP panels is accelerating the discovery of specific genes and gene areas and reinforcing the hunt for mutations driving underlying genetic variation in basic and complex traits. By conducting whole-genome association analyses, a significant number of SNPs have been found in domestic animals and livestock (particularly in bovine, porcine, and ovine species) in the last few years. Currently, commercially available whole-genome SNP array contains up to 750,000 SNPs. It is anticipated that this inexhaustible knowledge will help researchers for better understanding of the molecular mechanisms underlying the features they are interested in, including livestock growth, nutrition, general health and disease resistance. As a result, genomics suggests further ways to increase selection accuracy while lowering costs, shortening generation intervals and utilising novel sources of polymorphism.

Genomic Selection

Genomic selection refers to the process of choosing animals for breeding based on their genotype for tens of thousands of random SNPs that cover the entire genome. This process is also known as selective breeding. It is a sort of marker-assisted selection in which genetic markers covering the entire genome are utilised in order to ensure that all quantitative trait loci (QTL) are in linkage disequilibrium with at least one marker. This ensures that the best possible offspring will be produced. Genomic selection refers to the practise of making selection decisions by analysing genetic breeding values (GEBV). The GEBV are computed by summing the effects of dense genetic markers, or haplotypes of these markers, that are spread out across the entirety of the genome. This has the ability to capture all quantitative trait loci (QTL) that are responsible for the variation in the trait. In order to evaluate the impacts of QTL, phenotypic data from a large reference population are combined with either haplotypes or specific single nucleotide polymorphism markers. In later generations, the calculation of GEBV can be done with just the knowledge about the markers. These studies have the ability to detect statistical associations between commercially important traits and one or more SNP markers, which lends support to the development of comprehensive marker arrays for genomic selection. Genomic selection is a form of MAS that makes use of genetic markers that cover the entire genome. Utilizing the information provided by a large number of markers in order to evaluate breeding value without having prior knowledge of which specific genes are representative is one of the fundamental principles underlying genomic selection. Tens of thousands of single nucleotide polymorphisms (SNPs) are strategically placed across the entire genome in such a way that there is always an SNP in close proximity to a certain gene or DNA fragment of interest.

Potential of Genomic Selection

Improved selective breeding is one of most significant economic benefits of genomic selection. To estimate an improved EBV, the data on DNA markers, phenotypes, and pedigrees of possible selection candidates might be merged. This “genome-wide” selection’s EBV has a similar level of resilience to a traditional EBV that is based on the examination of pedigree and phenotype data. Selection decisions would be possible without phenotypic information once the association between chromosomal segments and phenotypes is established. The use of genomic selection is particularly appealing for attributes that are expensive to quantify. Additionally, it can be used more effectively for traits with low heritability, gradual genetic change, and measurement difficulties. The most likely use of this strategy will be in dairy cattle where, in theory, bulls might be chosen solely based on genomic data and utilised for mating long before the results of their progeny tests would be available, significantly shortening the generation interval and accelerating genetic progress.

Genomic Selection Produce Rapid Genetic Progress

There are three doable approaches to speeding up genetic development: improving selection accuracy, intensifying selection and shortening generation intervals, which also leads to more selection steps per unit of time. One of the few methods that can influence all three of these factors influencing genetic progress is whole-genome selection. Clearly, employing SNPs for genomic selection increases accuracy. Because SNP evaluations of embryos are just as valid as evaluations of young or elderly animals, the generation interval can be easily reduced. The method offers practical prospects for boosting selection intensity as well because it may be used widely and at a reasonable cost (screening hundreds of samples). The combination of these advantages, when added to pedigree and phenotypic information on each respective individual, becomes the most powerful, practical approach available for making genetic changes.

Reliabilities/Accuracy of Genomic Selection

For young bulls lacking progeny test results in the reference population, GEBV reliability ranged from 20% to 67%. Factors that affected the level of reliability attained included the heritability of the trait under exploration, the number of bulls in the reference population, the statistical method used to calculate the consequences of single nucleotide polymorphism in the reference population, and the reliability calculation technique. The precision of GEBV relies on several factors, including the size of the reference population, the multitude of animals with phenotypes and genotypes with in reference population from where the SNP effects are calculated, the extent of linkage disequilibrium (LD) among the markers and the QTL, the statistical approach used to estimate the SNP impact in reference population, the heritability of the trait to be evaluated, and the distribution of QTL effects. There is no way for breeders to influence how widely a trait is heritable or where a given QTL's effects are expressed. Genomic prediction in dairy cattle has a 0.8 accuracies for productivity traits and a 0.7 accuracies for fertility and lifespan.

Statistical Methods for Determining SNP Effects

Least square method tests all the genes one by one for their statistical significance and sets the effects of non-significant genes to zero, while estimating effect of significant genes. Best linear unbiased prediction (BLUP) fits the allelic effect as random instead of fixed. The fitting of random effect does not require a degree of freedom, thus all the allelic effects can be estimated simultaneously. Bayesian elimination is similar to BLUP except variance of allelic effect is assumed to be different for every marker and estimated using a prior distribution for the variance. Among these three, accuracy of genomic selection is highest in Bayesian elimination (0.84) followed by BLUP (0.74) and least square method (0.36). G-BLUP is simple extension of polygenic BLUP and is specifically efficient for polygenic traits. In G-BLUP all markers have same weight. Ridge regression- best linear unbiased prediction (RR-BLUP) assumes that all marker effects are normally distributed and have identical variance. It also assumes SNP effects 'm' random. Bayes A method assumes different variances of markers and normal distribution of each effect. Advance statistical methods for estimating SNP effects are G-BLUP, RR-BLUP, Bayes A (Meuwissen et al. 2001), Bayes C π (Legarra et al. 2001) and RR-BLUP B (modified RR-BLUP). Tools for SNP data analysis are AsReml (Gilmour et. al. 2006), PLINK (Purcell et. al 2007), Gensel (Rohan and Dorian 2009), GS3 (Legarra et al. 2011), solGS (Tecle et. al 2014), GVC BLUP (Wang et. al. 2014) and BLUPF 90 (Misztal et.al. 2015).

Difference between TAB, MAS and GS

The effectiveness of these two selection methods is affected by their primary distinction: scale. Since MAS relies so heavily on a small number of QTLs that are annotated by markers with strong associations, it can only predict so much about breeding values. Each QTL is guaranteed to be in LD with at least one SNP marker by utilising a large panel of markers from all around the genome, as is the case with GS. This gives GS a huge advantage over TAB and two major benefits over MAS. The complete heritability of a variable can be determined (i.e., it is hypothesis-independent in its approach). It is now possible to assess the population-level allelic effect of an SNP by capitalising on LD between the SNP and the QTL (no longer restricted to the family level).

Comparison of TAB, MAS, and GS Breeding Approaches

Features	TAB	MAS	GS
Breeding value based on animal phenotype scores (qualitative or quantitative)	✓		
Breeding value based on a combination of genetic-genomic data and phenotype scores		✓	✓
Selection of complex inheritance traits		✓	✓
Selection of production traits without negatively impacting other desired traits		✓	✓
Unlimited number of traits can be assessed		✓	✓
Accurate measurement of low heritability traits		✓	✓
Reproducible and inexpensive genetic assessment of large numbers of animals		✓	✓
Accurate genetic selection in both genders		✓	✓

Reduced generation interval for BV evaluation		✓	✓
Increased rate of genetic improvement in the animal population		✓	✓
Reduced cost for progeny testing		✓	✓
Selection of genomic markers across the entire genome			✓
Hypothesis-independent approach to tracking of all genetic variance			✓
Estimation of marker allelic effects for a population (not restricted to the family level)			✓

Genomic Selection in Different Livestock Species

Genomic Selection in Cattle

Besides having knowledge of DNA markers in dairy industry for decades (Smith, 1967; Soller and Beckman, 1983), use of DNA markers to improve the rate of genetic gain was very limited till 20th century. This was due to a number of factors. There are a huge number of loci that have an influence on several quantitative features, such as productivity and health traits in dairy cattle. However, each locus is only responsible for a small part of the total genetic variance. (e.g., Shrimpton and Robertson, 1988; Hayes and Goddard, 2001; Sanna et al., 2008, VanRaden et al., 2009). As a result of the limited number of markers available, only very minor increases in genetic gain were feasible, and the cost of genotyping these markers was expensive. The intricacy of calculating breeding values with marker information was another impediment to the use of marker assisted selection.

The first idea of predicting genetic merit from dense gene markers was given by Meuwissen et al. (2001) with 80% accuracy. Genomic selection using high-density SNP panels was first implemented in dairy cattle breeding programs. With advent of genotyping technology, genome sequences for human in 2001 and bovine in 2006, have allowed rapid detection of Single Nucleotide Polymorphisms (SNPs) and development of SNP-chips for rapid genotyping. Initially, A bovine SNP chip was developed for 10,000 (10k) markers and in 2007, a 56k chip was released by Illumina. Currently, the largest cattle chip available is 'IndiGau' which has 11496 markers (SNPs) placed on 777k Illumina chips of US and UK breeds. This chip was launched by National Institute of Animal Biotechnology (NIAB), Hyderabad. It is India's first cattle genomic chip for the conservation of pure varieties of indigenous cattle breeds like Gir, Kankrej, Sahiwal, Ongole etc.

Genomic selection gave advantages in terms of shorter generation intervals, increased accuracy in selecting young animals, and cost savings connected with traditional breeding operations' demand for progeny testing bulls (Hayes et al., 2009). Depending on the species of interest, the goal of dense arrays for

SNP detection has varying consequences. It has been used in farmed animals to identify markers that will boost the dependability of features linked to milk output, animal health, and animal conformation. This has been accomplished through the application of this technology. The discovery of genetic regions associated to milk production places Australia at the forefront among all nations (Bolormaa et al., 2010). Several studies have been conducted in a variety of nations, including the United States of America (Wiggans et al., 2009), Canada (Wiggans et al., 2009), China (Jiang et al., 2010), and others. The utilisation of genomic estimates in dairy cow breeding has led to significant changes, and the reliability of genomic prediction in dairy cattle now exceeds 0.8 for production traits and 0.7 for fertility and other variables respectively. Genomic selection is done on the sire side the majority of the time because it accounts for half of the herd and there is a huge availability of bull sperm for artificial intelligence breeding. Genomic selection allows for the early selection of superior bulls, which enables this process to begin at a younger age. When compared to genotyping heifer calves, genotyping young bulls increased genetic gain while also dramatically lowering costs. The decreased cost of genetic selection compared to traditional approaches has had a tremendous influence on the development of the dairy sector. Because of the high reliability and low cost of genome prediction from low-density genotyping followed by imputation, a significant number of selection candidates have been genotyped.

Because the generational gap between beef cattle and dairy cattle is so great, genomic selection has a much greater effect on the former. This means that genomic selection has the potential to dramatically increase genetic gain by bolstering selection reliability at a young age (Jonas & de Koning, 2015). However, the features of interest, such as growth rate, carcass, reproduction, and health, which all contribute to profitability, often cause indices to be selected based on a given market, delaying adoption (Montaldo et al., 2012). Genomic selection efficiency in beef cattle remains lower than in dairy cattle, perhaps because of breed heterogeneity, a lack of well-established structures and breeding programmes, the prevalence of natural service, cross-breeding in commercial herds, and smaller effective population sizes (Johnston et al., 2012). When compared to dairy cattle, beef cattle have less reliable genetic prediction (Van Eenennaam et al., 2014). The inferior quality and quantity of the beef cow population compared to the dairy cattle population is to blame for the diminished reliability. In addition, unlike dairy cattle, the aim population and validation animals in beef cattle may have a less genetic relationship to the reference population. To combat the problem of small reference populations, combining data from multiple countries and/or breeds may increase forecast accuracy (de Roos et al., 2009). The efficiency of genomic selection in beef cattle may be improved by better collection of genotypic data and phenotyping data.

Genomic Selection in Sheep

Genomic selection in the sheep was started with development and release of 57k ovine chip in 2008, which was based on a virtual ovine genome sequence (Dalrymple et al. 2007). The benefit of GS may be less in sheep because: 1) many traits under selection in sheep can be measured on both sexes prior to selection of animals for first mating; 2) several important traits have a high heritability; and 3) there may be less information available to derive accurate prediction equations due to fewer progeny tested sires. However, several sheep traits, including female fertility, slaughter traits, wool traits when assessed on adults, and parasite resistance, are difficult to measure on breeding animals. The advantage of GS is highest for traits that are difficult to quantify and have low heritability. The greatest advantage is a doubling of genetic gain. It should be highlighted that achieving high GS accuracy for characteristics with low heritability is more challenging in reality since considerably more phenotypic data is required to build GS prediction equations of such precision. Obviously, the advantage is greatest for features that cannot be assessed at all, unless such traits have extremely strong correlations with measured traits.

Genomic Selection in Pigs

The sequencing of the pig genome (Ramos et al., 2009) and the creation of the first commercial SNP panel for high throughput genotyping were the initial steps towards using genomic selection to pigs (Groenen et al. 2012). PorcineSNP60 BeadChip v2, at present available in version 2; Illumina, San Diego, CA) is the first commercially available SNP panel from Illumina, and it comprises around 60K SNPs that cover all autosomal and X chromosomes (Ramos et al. 2009). Several research have also proposed and simulated LD SNP panels with the aim of lowering genotyping costs, which are now one of the primary challenges preventing the widespread implementation of genomic selection in pigs (Habier et al. 2009; Dekkers et al. 2011; Wellmann et al. 2013). GeneSeek/Neogen (Lincoln, NE) designed the GeneSeek/Neogen GPP-Porcine LD Illumina Bead Chip Panel, a commercial LD SNP chip, to fill a gap in the market. Additionally, GeneSeek/Neogen developed a high-density SNP panel consisting of around 70K SNPs. High-density SNP (SNP) panel, covering all SNPs in a few genes linked to productivity attributes.

Primarily, pig genomic selection focuses on two categories of traits: maternal and performance. Ovulation rate, embryonic survival, measures of litter size (including overall number of piglets born, number of piglets born alive, number of weaned piglets, number of fully formed pigs, stillborn, pre-weaning mortality, and piglet survival), and piglet birth weight are all examples of maternal traits (low heritability traits). Genomic predictions for litter size and other maternal qualities in pigs generally beat pedigree-based forecasts, according to multiple genomic selection studies using either simulated or real data. Different maternal traits under genomic selection are total numbers of piglet born/litter size/total born (Cleveland et al. 2010; Forni et al. 2011; Jafarikia et al. 2012; Cleveland and Hickey 2013; Tusell et al. 2013; Uimari et al. 2014; Andersen-Ranberg and Grindfleck 2014; Hidalgo et al. 2015; Park et al. 2015), Number of born alive (Akanno et al. 2013, 2014), Stillborn, piglet mortality, litter weight (Andersen-Ranberg and Grindfleck 2014), gestation length (Hidalgo et al. 2015) and mortality rate (Guo et al. 2015).

Both performance and carcass variables have a moderate to high heritability, and specific breeding programmes have been developed in pigs in order to collect important phenotypes (on the same alive animals, after their slaughter, or on relatives like in the sib testing programs). Genomic selection is beneficial for performance traits due to the fact that phenotypic information can be leveraged across all selection candidates and potentially generations. Additionally, the collection of phenotypic data on a routine basis may be reduced, and there is a high degree of flexibility regarding the pigs that are to be phenotyped. This is an advantage for genomic selection (Dekkers, 2010). The important performance traits taken under consideration of genomic selection are daily gain/average daily gain (Ostersen et al. 2011; Christensen et al. 2012; Su et al. 2012; Akanno et al. 2013, 2014; Jiao et al. 2014), back fat thickness (Tribout et al. 2012; Akanno et al. 2013, 2014; Badke et al. 2014; Jiao et al. 2014; Do et al. 2015; Azevedo et al. 2015), feed intake/daily feed intake (Jiao et al. 2014; Do et al. 2015) feed efficiency (Tribout et al. 2012), feed conversion rate (Christensen et al. 2012), fattening trait (Tribout et al. 2013), number of days to 250 lb and loin muscle area (Badke et al. 2014), muscle depth and intramuscular fat content (Jiao et al. 2014), 45-min post-mortem carcass temperature and loin muscle pH (Ou et al. 2015) and bacon depth (Azevedo et al. 2015). Resistance to porcine reproductive and respiratory syndrome (Jafarikia and Sullivan 2014) and boar taint (Duijvesteijn et al. 2010) were also selected using genomic selection.

Genomic Selection in Poultry

Poultry breeding is very different from dairy cow breeding. Several distinct features determine how genomic selection may be employed for genetic improvement in chicken breeding. Poultry already has short generation intervals (broilers have multiple overlapping generations per year with selection every 6 weeks, while layers have non-overlapping annual generations). There is some potential for reducing the generation interval in layers, but not as much as in dairy cow breeding, where bull pathways might be lowered from more than 6 years to less than 3 years (Schaeffer, 2006, Schefers and Weigel, 2012). Large numbers of selection candidates and high selection intensities, along with low marginal revenues from a single selection candidate; tens of thousands of selection candidates are created per line every generation in chicken breeding programs, with just 1 to 3% of males selected for breeding. For genomic selection to be economically viable, genotyping needs to be relatively cheaper. Genetic progress in pure lines is propagated in poultry through a robust crossbreeding multiplication pyramid that affects a significant number of commercial layers and broilers. Thus, a single line contributes just 25% of the genes in commercial poultry compared to 100% in purebred dairy cow production systems. In comparison to dairy cattle, where large numbers of doses of semen are available from the top bulls due to large volume of semen produced by a bull and the cryopreservation facilities, highly selected males in poultry can be utilised to inseminate a limited number of birds. Individual contributions to improved commercial performance in the poultry industry are difficult to obtain since there is no pedigree information on commercial offspring of pure lines. The dairy sector, on the other hand, maintains pedigreed individual performance data on a significant number of commercial cows. Prior to implementing genomic selection in poultry, these aspects must be carefully considered.

In poultry, genomic selection has been routinely implemented after multiple well designed multi-generational selection trials in both layers and broilers (Misztal et al., 2013; Heidaritabar et al., 2014; Wolc et al., 2015). These trials validated genomic selection's claims of improved precision, higher responsiveness to selection and potential to reorganise breeding programmes to optimise the benefits of this technology. These preliminary studies also taught us about the practical use of genomics-based breeding strategies, such as large-scale genotyping with adequate sample monitoring. The development of high-density SNP chips, which allow for quick, vast, and relatively affordable genotyping, is a prerequisite for genomic selection. The chicken was the first livestock species to be sequenced (Hillier et al., 2004), and many million SNPs were discovered at the same time (Wong et al., 2004). The initial chicken SNP chip had just 3,000 (3K) SNPs (Muir et al., 2008), which was inadequate for the genomic selection. Two medium-density chips, 60K chip (developed and used by the Cobb and Hendrix groups in collaboration with USDA) and 42K chip (developed by the EW group) were designed to include genetic variation in both broiler and layer lines. Further, a high-density 600K chip was developed with support from the BBSRC and the Roslin Institute (Kranis et al., 2013), which comprises variants from both brown and white egg layer breeds and from broiler breeds. Currently, medium density SNP chips containing tens of thousands of SNPs are used over high-density density chips for routine implementation of genomic selection in poultry because of the lower cost of medium-density panels due to technological advancements, the stabilised accuracy of prediction of genomic breeding values within a line or breed and the simplified genomic prediction workflow.

Genomic selection in a line of layers with commercial significance was initiated for the first time by Hy-Line International in 2013, and it was preceded by three years of genomic selection in an experimental line (Wolc et al., 2015). Initial implementation involved HD genotyping (600K panel) of several generations of selected males and granddams of selection candidates, followed by low-density genotyping (1,000 SNPs, in-house genotyping) of selection candidates and their dams, with subsequent imputation to 600K. Accuracy changes for different approaches varied from negative (for complex multitrait models with genomic information that failed to converge) to a 40% improvement over pedigree-BLUP for BayesB with p = 0.99. Because of the increased amount of training data, the average accuracy of GEBVs increased somewhat in each successive generation. Other layer breeding companies have also conducted genomic selection studies in their populations, with promising outcomes (Sitzenstock et al., 2013; Visscher, 2015).

Aviagen and Cobb Vantress, two prominent breeding companies, conducted research on the use of genomic selection in broilers. With broiler generation intervals already quite short, the main impact of genomic selection is on EBV accuracy, particularly for features that are not accessible on selection candidates at the time of selection. These characteristics include reproductive performance and disease resistance. Considerably, when selection candidates were genotyped using fewer than 400 markers and imputed up to 42K, the imputed genotype accuracy was in the 97% range (Wang et al., 2013), with findings for the sex chromosomes being even higher (Hickey and Kranis, 2013). Increasing the number of SNPs in the low-density array to 3K further increased the accuracy of imputation up to 600K to 99% (Wolc et al., 2014).

Conclusion

Poor breeding infrastructure, a lack of systematic recording of reliable phenotypes, and appropriate analytical tools to synthesise the data, providing prompt feedback to aid improve farmer management and husbandry procedures, are the key limiting issues for the adoption of GS in underdeveloped countries. There are still significant obstacles and possibilities in applying genomic selection, such as adapting national genetic assessments to integrate genomic information, genomic selection across breeds, controlling long-term gain and inbreeding with genomic selection, and computing challenges. Genomic selection provides two opportunities, that have not been fully utilised yet. First, combining GS with reproductive technology has the potential to greatly shorten generation time; for example, using GS to select the best embryos to produce the next generation of animals through multiple ovulation and embryo transfer (MOET) (rather than random embryos). Second, rather than stud animals that have been tested for economically important qualities, we may train the prediction equation on commercial animals. The benefits of genomic selection are smaller in small ruminants. In case of pigs, genomic selection is more efficient for maternal traits (low heritability) in comparison to performance traits. The developing domains of pig metabolomics and phenomics may provide a rich potential source of novel and intriguing features related to pig production level and quality, health status, and well-being. GEBV prediction may be explored for the most relevant characteristics, both in the context of selection for the trait in question and as a signal of the pigs' status, given that almost all new phenotypes are costly to assess and can only be obtained on a small number of animals. Despite the fact that genomic selection has been used in some layer and broiler breeding projects, it is still a fast-emerging field. With recent advances in sequencing, other poultry species are expected to be the next to undergo genomic selection. New models, such as Bayesian single step, haplotype and QTL models and non-additive models are now being tested in poultry data.

References

Boichard D, Ducrocq V, Croiseau P and Fritz S. 2016. Genomic selection in domestic animals: Principles, applications and perspectives. ComptesRendusBiologies. 339: 274–277.

Calus M P L, Meuwissen T H E, de RoosA P W and Veerkamp R F. 2008. Accuracy of Genomic Selection Using Different Methods to Define Haplotypes. Genetics. 178: 553–561

Daetwyler H D, Villanueva B, Bijma P and Woolliams J A 2007. Inbreeding in genome-wide selection. Journal of Animal Breeding and Genetics. 124:369–376.

Fulton J E. 2012. Genomic Selection in poultry breeding. Animal Frontiers. 2(1): 30-36

Goddard, M E and Hayes B J. 2007. Genomic selection. Journal of Animal Breeding and Genetics. 124:323–330.

Hayes B J, Bowman P J, Chamberlain A J and Goddard M E. 2009. Genomic selection in dairy cattle: Progress and challenges. Jornal of Dairy Science. 92:433–443

Meuwissen T H E, Hayes B J and Goddard M E. 2001. Prediction of total genetic value using genome-wide dense marker maps. Genetics. 157:1819–1829.

Meuwissen T, Hayes B and Goddard M. 2016. Genomic selection: A paradigm shift in animal breeding. Animal Frontiers. 6(1): 6-14

Morde R, Ojango J M K, Okeyo A M and Mwacharo J M. 2019. Genomic selection and use of molecular tools in breeding programs for indigenous and crossbred cattle in developing countries: Current status and future prospects. Frontier in Genetics.

Samore A B and Fontanesi L. 2016. Genomic selection in pigs: state of the art and perspectives. Italian Journal of Animal Science. 15(2): 211-232

Van der Werf J H J. 2008. Potential benefit of genomic selection in sheep. Proc. Assoc. Advmt. Anim. Breed. Genet. 18:38-41

Weigel K A. 2017. Genomic selection of dairy cattle: A review of methods, strategies, and impact. Journal of Animal Breeding and Genomics. 1(1): 1-15

Wolc A, Kranis A, Arango J, Settar P, Fulton J E O'Sullivan N P, Avendano A, Watson K A, de los J M G Campos,Fernando R L, Garrick D J and J.C.M. Dekkers. 2016. Implementation of genomic selection in the poultry industry. Animal Frontiers. 6(1): 23-31

17

Biotechnological Interventions for Enhancing Wool Production

Dibyendu Chakraborty, Komal Deep Kour, Ishani Sharma and Harpreet Singh

Division of Animal Genetics & Breeding, FVSc&AH, SKUAST-Jammu, R.S. Pura Jammu, UT of J&K, India

Introduction

Wool which obtained primarily from sheep is an essential textile fiber that is used across the world. Whether for clothing, canvas work, carpets, or even upholstery, wool is a vital resource supporting many industries and occupations. This natural fiber has many benefits, such as being biodegradable, easy to clean and even having a natural structure that traps air which is excellent for heat retention. The annual wool production is 2 million tones. China ranks first in terms of global wool production. With a production of more than 333,624 tonnes of wool in 2020, China accounts for approximately 19% of the global wool clip. Australia produces 16% of global wool and is by far the largest wool exporter in the world. There has been a constant increase in the demand of wool products worldwide. This has made scientists to try to improve livestock and livestock associated derivatives. The conventional system along with the recent biotechnological interventions will be helpful for enhancing wool quality and production.

Biotechnological Interventions

With genetic manipulation and related technologies gaining prominence more and more, research interests to improve livestock using genetic engineering has increased tremendously (Onteru *et al.*, 2010). Genetic engineering refers to the direct manipulation of an organism's genome using modern DNA technology that involves integration of foreign DNA or synthetic genes into the organism of interest, necessary gene manipulation for desired output and a desired change in genotype or phenotype. Genetic engineering falls within the ambit of modern biotechnology (Robl *et al.*, 2007). Wool production is highly heritable; however, nutrition can affect wool growth and character. A deficiency of Selenium (Se) depressed wool production by lambs by 35%. There are several studies suggested that by providing additional Se, wool production has been increased in both ewes and lambs. Therefore, the biotechnological interventions both in sheep genome level as well as microbial and forage levels would be beneficial for wool production and quality improvement.

There are three principal approaches for genetic manipulation;

(a) to introduce new genes or modification of gene/gene sequences into the sheep's genome

(b) to introduce new genes into one of the ruminal microorganisms

(c) to engineer forage plants which make proteins which can withstand ruminal breakdown. Transgenic sheep carrying a keratin-IGF-I construct expressed in their skin produced 6.2% more clear fleece than non-transgenic controls and no adverse effects on health or reproduction were observed (Damak *et al.*, 1996a; Damak *et al.*, 1996b). Since cystein is the rate-limiting biochemical attribute for wool growth, similar attempts have been undertaken to change wool production through transgenic modification of the cystein pathway (Ward and Armstrong, 2000).

CRISPR-Cas

Clustered regularly interspaced short palindromic repeat (CRISPR)/CRISPR-associated 9 (Cas9) to generate loss-of-function mutations with the *FGF5* gene in Chinese Merino sheep showed that wool staple length and stretched length of genetically modified (GM) yearling sheep were significantly longer compared with that of wild-type (WT) control animals. The greasy fleece weight of GM yearling sheep was also significantly greater compared with that of WT sheep. There was no significant difference in the mean fiber diameter in GM sheep compared with WT sheep. The results suggested that CRISPR/Cas9-mediated loss of FGF5 activity could promote wool growth and, consequently, increase wool length and yield (Li *et al.*, 2017).

Transgenic Sheep with Insulin Like Growth Factor 1

Different methods for generating transgenic livestock include microinjection (Gordon *et al.*, 1980), sperm mediated DNA transfer (Smith and Spadafora, 2005), intracytoplasmic injection (ICSI) of sperm heads carrying foreign DNA injection or infection of oocytes (Li *et al.*, 2012) and/or embryos by different types of viral vectors (Ishii *et al.*, 2004), RNA interference technology (RNAi) (Wise *et al.*, 2008) , use of somatic cell nuclear transfer (SCNT) (Mir *et al.*, 2005) and using transposons as vectors (Peng *et al.*, 2010). By pronuclear microinjection with a mouse ultra-high-sulfur keratin promoter connected to an ovine insulin-like growth factor 1 (IGF1) cDNA, transgenic sheep were created. The transgene consisted of a mouse keratin promoter (McNab *et al.*, 1990) linked to a sheep IGF-1 cDNA (Wong *et al.*, 1989). Transgenic animals were identified by PCR, using primers specific for the keratin promoter and the IGF-1 cDNA, respectively. Transgenic sheep carrying a keratin-IGF-I construct expressed in their skin produced 6.2% more clear fleece than non-transgenic controls and no adverse effects on health or reproduction were observed (Damak *et al.*, 1996a; Damak *et al.*, 1996b). The rate of fiber growth is determined by the proliferation of epithelial cells in the follicle bulb and their subsequent migration and growth (Chapman *et al.*, 1980). Local infusion of IGF-1 in the skin of sheep has been shown to increase local blood flow, oxygen utilization and amino acid uptake but not to increase cell proliferation (Harris *et al.*, 1993). Thus it is likely that the effect is due to stimulation of follicle metabolism by IGF-1. The keratogenous zone of the wool follicle cortex was the only place where the transgene was expressed in the skin.

Insertion of Biochemical Pathways by Transgenesis of Sheep

Metabolism and degradation of the sulphur, containing amino acids, cysteine and methionine, by microorganisms of the sheep rumen lead to loss of much of the dietary sulphur as sulphide. This is known to limit the growth of wool, which is composed of many cysteine-rich proteins (Powell and Rogers, 1990). Infusion of methionine or cysteine either directly into the bloodstream or into the abomasum (the glandular compartment of the ruminant stomach) of sheep leads to an increase in the rate of mitosis (Hynd, 1989) and wool growth (Reis *et al.*, 1990). One way to overcome this loss would be to produce transgenic sheep containing bacterial genes for cysteine synthesis. These genes would be modified such that their expression in the epithelial walls of the upper gastrointestinal tract would enable the liberated sulphide to be 'trapped'. The two genes were isolated from *Salmonella typhimurium*, sequenced, separately fused to the SV40 late promoter and polyadenylation signal in a plasmid vector 8'9, and expressed in sheep cells in culture, producing active serine acetyltransferase (SAT) and O-acetylserine sulphydrylase (OAS) (Cherest *et al.*, 1993). In bacteria, the cysteine biosynthesis pathway has been characterized by two-step pathway of conversion of serine to cysteine via an intermediate, O-acetyl serine (OAS) is catalysed by the enzymes (i) serine acetyltransferase (SAT; encoded by the cysE gene) and (ii) Oacetyl serine sulphydrylase (OASS-A encoded by the cysK gene and OASS-B by the cysM gene). Coexpression of genes encoding the bacterial cysteine biosynthesis pathway indicated that it is possible to coexpress linked cysE and cysM genes in mammalian cells to produce cysteine (Bawden *et al.*, 1991). In another laboratory the genes from *E. coli* cell encoding SAT and OAS are being fused with the promoter from the sheep metallothionein gene Ia (MT-Ia) (Ward *et al.*, 1989). This promoter had previously been fused with the gene for ovine growth hormone, in order to gain external control by dietary zinc.

The two genes for the glyoxalate pathway, aceA and aceB, have been isolated from *E. coli* and inserted into a plasmid under the control of promoters that function in animal cells. Glucose produced in this way not only reduce the use of amino acids but also supply the energy to follicle needed for wool growth (Chapman *et al.*, 1980) .

Modifying the Protein Composition of Wool

The physical properties of wool fibre can be attributed to proteins from the keratin family which are the primary constituents of the fibre (Itenge-Mwezaa *et al.,* 2007). The two main proteins that make up the fibre of wool are keratin-associated proteins (KAPs) and keratin intermediate-filament proteins (KRTs). The wool fiber's microfibrils' skeletal structure is formed by the KRTs, which are enmeshed in a matrix of KAPs. KRT1.2 polymorphism by PCR-RFLP was found to be associated with different wool parameters like GFW, FD and SL (Singh *et al.*, 2022).

It is clearly possible to transfer active wool fibre protein transgenes into the genome of existing sheep and genes encoding Keratin Intermediate Filaments (KIF) or Keratin-Associated Proteins (KAPs) i.e KAP4 and KAP6 may be used. It has been found that using a well characterised wool follicle cortex-specific promoter, transgenes added to the sheep are active in the desired cortical cell compartment. Transgene mRNA is made and that proteins encoded by the transgene are synthesised. Wool fibre macrostructure (e.g. fibre crimp), ultrastructure (i.e. the ordered KIF-KAP arrangement), other fibre features (e.g. fibre lustre), physical properties, chemical composition and structural features of the modified wool fibres depends upon the level of expression of the transgene, and the timing and location of transgene protein synthesis. (Bawden , 2009)

Engineering Rumen Bacteria

Introduction of diverse genes in gut micro organisms enhances their capacity of to digest fibrous components and lignins of forage, or degrade toxins, synthesize essential amino acids, reduce ruminal methane production and tolerate acids . Introducing new species or strains of microorganisms in rumen increase digestibility of feedstuffs and improve animal production (Ulfina *et al.*, 2019). First step in the process of genetic modification of rumen microbes is the selection of desired gene which has to be engineered. After selecting the desired gene a suitable vector for carrying the gene to the recipient cell is required. One of the most important vector is plasmid. Thus the genetic manipulation/engineering work can be done in bacteria, easy to handle e.g. *Escherichia coli* and then transfer the recombined genetic material into the rumen bacteria which can be used for practical application. The genes which have been cloned in *Escherichia coli* are endoglucanase, xylanase, β-glucosidase, amylase, glutamine synthetase from the donor source of *Bacteroides fibrisolvens, Ruminococcus flavefaciens, Fibrobacter succinogenes, Neocallimastix frontalis, Streptococcus bovis* etc (Santra and Karim, 2003).

Engineering Forage Plants

The genetic manipulation of forage plants is another way in which an improved balance of amino acids, including cysteine, might be supplied for wool growth (Roger,1990). Several criteria would have to be met. First, a gene which encodes a protein with the appropriate amino acid composition is required. Second, the protein must be relatively resistant to ruminal degradation (to prevent the loss of amino acids to rumen microflora) third the gene must be highly expressed in the leaves and stems of pasture species such as lucerne and subterranean clover. A number of candidate proteins from several seeds have been examined. A low-molecular-weight pea seed-protein (pea albumin 1 or PAl) rich in cystine was found to be resistant to rumen fluid both in vitro (Spencer *et al.*, 1988) and in situ and the gene for this protein has been isolated and sequenced (Higgins *et al.*, 1986). The genes for PAl and for another pea seed protein, vicilin, have been introduced into tobacco test plants (*Nicotiana tabacum*) and into lucerne (*Medicago sativa*) utilizing the binary vector system derived from the Ti plasmid of *Agrobacterium tumefaciens*. Alternatively, techniques such as electroporation or 'biolistics', could be used to transfer DNA into the plant cells.

Conclusion

The prospects for applying recombinant DNA methods and other recent biotechnological tools to improve wool growth depend on the research resources available. Transgenesis in sheep, in rumen bacteria and in forage plants, are all major projects. The relatively high cost of experimenting with transgenic sheep means that careful planning is required. In principle, the ruminal bacterium approach has greater flexibility, since bacteria are easier and cheaper to manipulate and provide a wide diversity of organisms to work with. However, when procedures for animal transgenesis become more efficient and targeted, the range of ways in which wool growth could be improved might be extended: for instance, the effects of specific growth factors on follicle cell proliferation or the expression of specific molecules to prevent insect infestation or promote systemic resistance to disease might be examined. Likewise, improvements in plant transgenesis will see further development in fodder improvements. Without doubt, it has improved tremendously over the years, but the wool-grower ultimately stands to gain from the current research.

References

Bawden, C.S., Cam, G.R., D'Andrea, R.J., Kuczek, E.S., Powell, B.C., Rogers, G.E. and Sivaprasad, A.V. 1987. A review of the genes for woof keratin and for cysteine biosynthesis: their significance for studies directed towards modified wool production. In: McGuirk B.J. (ed.), Merino Improvement Programs in Australia. pp. 511-8.

Bawden .S .2009. Genetic Engineering. Wool Biology and Metrology. 4 :472-572.

Chapman, R.E., Downes,A.M. and Wilson,P.A. 1980. Migration and keratinization of cells in wool follicles. Aust. J. Biol. Sci., 33:587-603.

Cherest, H., Thomas, D. and Surdin-Kerjan, Y. 1993 Cysteine biosynthesis in Saccharomyces cerevisiae occurs through the transsulfuration pathway which has been built up by enzyme recruitment. J. Bacteriol., 175: 5366-74.

Damak ,S., Jay, N.P., Barrell, G.K. and Bullock, D.W. 1996a. Targeting gene expression to the wool follicle in transgenic sheep. Biotechnology. 14: 181-184.

Gordon. J.W., Scangos, G.A., Plotkin, D.J., Barbosa, J.A., and Ruddle, F.H. 1980. Genetic transformation of mouse embryos by microinjection of purified DNA. Proc Natl Acad Sci., 77: 7380-7384.

Harris, P.M., McBride,B.W., Gurnsey, M.P., Sinclair,B.R. and Lee,J. 1993. Direct infusion of a variant of insulin-like growth factor-I into the skin of sheep and effects on local blood flow, amino acid utilization and cell replication. J. Endocr., 139: 463-472.

Higgins, T. J. V., Chandler, P. M. and Randall, P. J. 1986. J. Biol. Chem., 261:11124-11130

Hynd, P.I. 1989. Effects of nutrition on wool follicle cell kinetics in sheep differing in efficiency of wool production. Aust. J. Agric. Res. 40: 409-17.

Ishii, Y., Reese, D.E. and Mikawa, T. 2004. Somatic transgenesis using retroviral vectors in the chicken embryo. Dev Dyn., 229: 630-642.

Itenge-Mwezaa, T. A., Forresta, R. H. J., McKenzie, G. W., Hogana, A., Abbott, J., Amoafoa, O. and Hickford, J. G. H. 2007. Polymorphism of KAP1.1, KAP1.3 and K33 genes in Merino sheep. Molecular and Cellular Probes, 21: 338-342.

Li,M.W., Baridon, B., Trainor, A., Djan, E., Koehne, A., Griffey, S.M., Biggers, J.D., Toner, M. and Lloyd, K.C. 2012. Mutant mice derived by ICSI of evaporatively dried spermatozoa exhibit expected phenotype. Reproduction,143: 449-453.

Li WR, Liu CX, Zhang XM, Chen L, Peng XR, He SG, Lin JP, Han B, Wang LQ, Huang JC, Liu MJ. CRISPR/Cas9-mediated loss of FGF5 function increases wool staple length in sheep. FEBS J. 2017 Sep;284(17):2764-2773. doi: 10.1111/febs.14144. Epub 2017 Jul 26. PMID: 28631368.

McNab, A.R., Andrus,P., Wagner,T.E., Buhl,A.E., Waldon, D.J., Kawabe, T.T., Rea,T.J., Groppi,V. and Vogeli,G. 1990. Hair-specific expression of chloramphenicol acetyl transferase in transgenic mice under the control of an ultra-high-sulfur keratin promoter. Proc. Natl. Acad. Sci. USA.,87: 6848-6852.

Mir, B., Zaunbrecher, G., Archer, G.S., Friend, T.H. and Piedrahita, J.A. 2005. Progeny of somatic cell nuclear transfer (SCNT) pig clones are phenotypically similar to non-cloned pigs. Cloning Stem Cells, 7: 119-125.

Onteru, S., Ampaire, A. and Rothschild, M. 2010. Biotechnology developments in the livestock sector in developing countries. Biotechnol Genet Eng Rev., 27: 217-228.

Peng, K.C., Pan, C.Y., Chou, H.N. and Chen, J.Y. 2010. Using an improved transposon system to produce transgenic zebrafish with epinecidin-1 which enhanced resistance to bacterial infection. Fish Shellfish Immunol., 28: 905-917.

Powell, B.C. and Rogers, G.E. 1990. Hard keratin IF and associated proteins. In: Goldman R.D. and Steinert P.M. (eds), Cellular and Molecular Biology of Intermediate Filaments 267-300.

Reis, P.J., Tunks, D.A. and Munro, S.G. 1990. Effects of infusion of amino acids into the abomasum of sheep, with emphasis on the relative value of methionine, cysteine and homocysteine for wool growth. J. Agric. Sci., Camb., 114: 59-68.

Robl, J.M., Wang, Z., Kasinathan, P. and Kuroiwa, Y. 2007. Transgenic animal production and animal biotechnology. Theriogenology. 67: 127-133.

Santra, A. and Karim, S. A. 2003. Rumen manipulation to improve animal productivity. Aasian-Australian Journal of Animal Sciences, 16(5): 748-763.

Singh, V. P., Taggar, R. K., Chakraborty, D., Singh, B. P., Khajuria, P., Singh, S. and Gupta, P. 2022. KRT 1.2 gene polymorphism & its association with wool traits in Rambouillet sheep. The Pharma Innovation Journal, SP-11(6): 2619-2621.

Smith, K. and Spadafora, C. 2005. Sperm-mediated gene transfer: applications and implications. Bioessays .27: 551-562.

Spencer, D., Dove, H. and Coombe; J. B. 1988. Brit. J. Nutr. 60: 241-247

Ulfina, G., Lemma, F., Tekalign, T. and Amanuel, B. 2019.Rumen Manipulation: One of the Promising Strategies to Improve Livestock Productivity-Review. Dairy and Vet Sci J., 9(2): 555758. DOI:10.19080/JDVS.2019.09.555758

Ward, K. A., Murray, J. D. and Nancarrow, C. D. 1989. Biotechnology for Livestock Production, FAO. Animal Production and Health Division, pp. 17-28.

Ward, M.P. and Armstrong, R.T. 2000. Inspection of wool lots at sales as a diagnostic test for louse infestation. Vet Parasitol., 90: 119-128.

Wise, T.G., Schafer, D.S., Lowenthal, J.W. and Doran, T.J. 2008. The use of RNAi and transgenics to develop viral disease resistant livestock. Dev Biol., 132: 377-382.

Wong, E.A., Ohlsen,S.M., Godfredson, J.A., Dean, D.M. and Wheaton,J.E. 1989. Cloning of ovine insulin-like growth factor-I cDNAs: heterogeneity in the mRNA population. DNA, 8: 649-657.

18

Genetic Improvement of Malpura Sheep Through Selective Breeding for Mutton Production

P.K. Mallick and S.S. Misra

Division of Animal Genetics and Breeding, ICAR-Central Sheep & Wool Research Institute,Avikanagar, Malpura (Tonk), Rajasthan

Livestock is one of the fastest growing agricultural subsectors in developing countries. In India has a vast diversity of animal genetic resources. Among them, sheep is an important 'five star' livestock species contributing significantly to rural economy through production of mutton, wool, manure, skin and milk etc. Sheep make a valuable contribution to the livelihood of the economically weaker sections of the society. The total Livestock population is 535.78 million and total sheep population is 74.26 million in in the country showing an increase of 4.6% and 14.1%, respectively over Livestock Census-2012. And it's ranking 3rd in the world after China and Australia. Rajasthan is the fourth largest sheep rearing state in India with 7.9 million population (Livestock Census, 2019). Malpura sheep is one amongst the heaviest sheep breed of India, widely distributed in the semi-arid region of Rajasthan, mostly Tonk, Jaipur and Sawai Madhopur districts. Malpura sheep is reared and improved for mutton production since year 1975 at CSWRI, Avikanagar. The objective since then has been to improve the growth characteristics of this sheep breed for producing more carcass yield.The sheep is known for its adaptability to the harsh environment and potential for high meat production.Coat colour of Malpura sheep is white with coarse wool. The face is light brown with short, blunt ears. Animals have square compact body and long legs. Farmers also have liking for the long tail of animals for aesthetic purpose. The farmers like Malpura sheep for its properties like good mothering ability and produce sufficient milk to sustain and accelerate the growth of their neonates (Mishra et al., 2005), higher weight gain, number of lambs weaned, wool quality, walking capacity and the ease of management.Malpura sheep is reared by small and marginal land holders who graze them on fallow land, crop residue and also take them on migration during period of scarcity of feed. The Central Sheep & Wool Research Institute (CSWRI) Avikanagar is involved in genetic improvement and conservation of Malpura sheep since many years. Animals are very well adapted to variations in climate and scarcity of the grazing resources as they are able to maintain the live weights across the year in the average range of 33 to 40 kg for females and 39 to 50 kg in males with significant variations between the seasons.Indian Council of Agricultural Research (ICAR) has started mega project for dissemination of improved germplasm (plants and animals) to the field during XI five year plan. Malpura sheep unit was inducted in the Mega Sheep Seed platform since March 2014. This unit was started at the ICAR-Central Sheep and Wool Research Institute Avikanagar, where already a well-established nucleus of genetically improved Malpura sheep was present. MSSP, Malpura sheep started with the aim to improving the growth production, changing the traditional management practiceand decrease the mortality rate for the livelihood of farmers in its breeding tract.

Management Practices Followed

Animals according to their age, sex, physiological and health status were housed in the sheep shed providing adequate floor space. All the sheds and corrals were disinfected frequently with lime. Sufficient grazing area was available for grazing but due to drought condition in the state grazing material availability was poor during the year especially during the lean period (March to July) and animals was provided supplementary feeding. Contractual grazing system is being followed. The grazing hours varied from 10 to 12 hours per day. Concentrate was supplemented to the pregnant, lactating, breeding rams and suckling lambs at the rate of 250 to 300 g per day. Various prophylactic measures viz. drenching, dipping and vaccination were followed. To maintain hygienic sheep sheds and corrals, six inches earth was scrapped and replaced with new earth mixed with lime. The sick animals were isolated from the flock and were treated accordingly. Heat detection was done in the morning and evening. The ewes were mated with the selected rams in the morning. Mating plan was designed in such a way so that inbreeding could be avoided.

Salient Achievements (present status vs inception)

- Presently (2020 born), the least squares mean (LSM) of weaning weight (3 month) is 17.91 kg which is about 97.68% improved over that of inception of the project (9.06 vs 17.91).

- The LSM of 6 and 9 month body weight are 26.52 and 29.33 kg, respectively, which are about 92.59% and 80.04% improved over the inception of the project (13.77 vs 26.52 and 16.29 vs 29.33 kg)
- The LSM of 12month body weight is 35.29 kg, which is 91.79% improved over the inception of the project (18.40 vs 35.29).
- The adult survivability was 96.59% which was improved by about 22.16% over the inception.
- The lambing percentage on tupped basis was 95.37% which was about 34.57% improved over that of inception.
- The wool yield and quality performance at present is improved over the inception by 46.69% (0.771 vs 1.131kg in adult annual GFW) and modulation % also decreases 17.96 (82.70 vs 67.85).
- The genetic trend of 6month body weight was upward and positive. The genetic gain was 25.3g per year. The high R^2 (0.696) indicates genetic selection was very effective in improving growth performance of Malpura Sheep.
- The inbreeding coefficient was 2.04% which was well within the acceptable limit.
- For the genetic improvement of the animals in the farmers flock and a total of 631 superior breeding rams/ram lambs were supplied/sold to the Govt. Agency/ MSSP Farmers/NGO/ Developmental agency during the reported period. Since 2014 onwards in Comparison of growth performance of progeny of distributed rams and farmer's rams, it shows that progeny of institute provided ram is always better than farmers ram.

Table 1: Summary of Progress report

S.No.	Particulars	At inception 1975	At present 2020	improvement % over inception
A. Growth performance (LSM)				
	Birth Wt (kg)	2.40	3.34	39.17
	3 M Wt (kg)	9.06	17.91	97.68
	6 M Wt (kg)	13.77	26.52	92.59
	9 M Wt (kg)	16.29	29.33	80.04
	12 M Wt (kg)	18.40	35.29	91.79
B. Wool yield & quality traits				
	1st six monthly GFW (g)	0.506	0.560	1.98
	Adult annual GFW (g)	0.771	1.131	46.69
	Fiber diameter (µm)	39.41	46.75	18.62
	Medullation (%)	82.70	67.85	17.96
	Staple length(cm)	5.76	5.56	-3.47
B. Survivability%				
	0-3M	91.21	92.34	1.24
	3-12M	92.14	97.36	5.67
	Adult	79.07	96.59	22.16
B. Reproductive traits				
	Tupping %	94.07	109.09	15.95
	On ewes tupped basis	70.87	95.37	34.57

Flock statistics: Malpura sheep were rearing since 1975 at CSWRI, Avikanagar for the different breed evolution programe. From the year 2014 Malpura sheep is maintaining at ICAR-CSWRI under the platform of Mega Sheep Seed Project the flock statistics is given since 2009 in Table 2.

Table 2: Nucleus flock strength (closing balance as on 31st March)

Particulars	2009-10 (31.03.2010)	2010-11	2011-12	2012-13	2013-14	2014-15	2015-16	2016-17	2017-18	2018-19	2019-20
Breedable female	431	465	434	330	238	274	316	337	396	442	453
Breeding male	109	75	144	86	73	86	65	73	78	54	72
Hogget (6-12 M)	38	145	45	24	28	0	05	59	8	102	175
Weaner (3-6 month)			87	1	3	77	125	02	91	12	9
Suckling lambs (0-3 Month)	224	107	136	266	137	118	185	165	182	119	179
Total	802	792	846	707	479	555	696	636	755	729	888

1. Growth of Malpura sheep

More than 8000 birth records of Malpura sheep has been selected for the growth traits at different stage over the period of 45 years (1975-2020) were analyses to observe the overall performance of animals at different stages and period. Growth performance of Malpura sheep during last period is tremendous improvement over the earlier performance setting a new insight in the growth potential of Malpura sheep and possibility of its use as a good mutton type sheep.

Table 3: Variability measurements for the growth Traits in Malpura sheep since inception 1975-2020

Total	Properties	BT	3WT	6WT	12WT
	Mean	3.07	14.72	22.45	29.09
	N	8656	7522	6562	4804
	Std. Deviation	0.62	3.91	5.52	6.24
	Range	5.00	27.20	40.00	46.00
	Variance	0.50	15.49	30.00	37.68
	CV	22.97	26.86	24.53	21.37

Table 4: Growth performance (LSM±SE) during the period 1975-2020

Effect/trait	Birth Wt.	3M Wt.	6M Wt.	9MWt	12M Wt.
Overall mean	3.06±0.02 (8798)	14.16±0.05 (7673)	21.50±0.07 (6670)	24.71±0.08 (5604)	28.12±0.09 (4871)
Period of birth	**	**	**	**	**
P1(1975-1979)	2.84±0.03 (369)	11.87±0.18 (311)	18.73±0.27 (258)	20.82±0.35 (168)	23.08±0.38 (154)
P2(1980-1984)	3.09±0.04 (235)	13.98±0.22 (209)	20.01±0.33 (170)	22.93±0.43 (112)	25.73±0.30 (99)
P3(1985-1989)	3.18±0.03 (443)	15.13±0.17 (347)	22.26±0.26 (269)	26.09±0.28 (252)	29.37±0.30 (224)
P4(1990-1994)	3.08±0.02 (639)	13.73±0.14 (506)	20.40±0.19 (496)	23.73±0.22 (426)	26.41±0.23 (382)
P5(1995-1999)	3.10±0.02 (806)	13.10±0.13 (638)	20.18±0.17 (641)	22.90±0.19 (570)	26.98±0.23 (421)
P6(2000-2004)	2.95±0.02 (930)	13.90±0.10 (882)	21.49±0.15 (843)	25.32±0.16 (722)	28.87±0.18 (620)
P7(2005-2009)	3.00±0.01 (1270)	14.16±0.09 (1100)	21.36±0.14 (1004)	24.77±0.16 (798)	28.99±0.17 (714)
P8(2010-2014)	3.14±0.02 (1561)	15.40±0.08 (1387)	23.65±0.13 (1102)	27.42±0.13 (948)	31.41±0.15 (875)
P9(2015-2020)	3.13±0.01 (2545)	16.19±0.08 (2293)	25.39±0.11 (1887)	28.39±0.12 (1608)	32.28±0.13 (1382)
Sex	**	**	**	**	**

Effect/trait	Birth Wt.	3M Wt.	6M Wt.	9MWt	12M Wt.
Male	3.13±0.01 (4381)	14.78±0.06 (3717)	23.21±0.09 (3208)	26.87±0.11 (2465)	30.97±0.12 (1998)
Female	2.99±0.01 (4417)	13.55±0.06 (3956)	19.79±0.08 (3462)	22.54±0.09 (3139)	25.87±0.10 (2873)
Reg Ewe Wt	**	**	**		**

Growth traits in Malpura sheep from the year 1975-2020

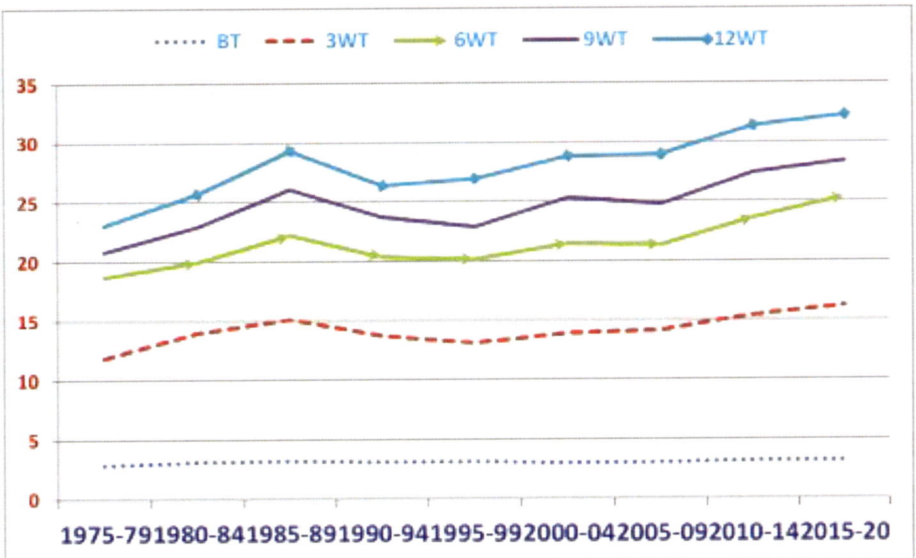

2. Greasy Fleece Yield

Regular and timely shearing of wool from the body provides comforts to the animals. The data on First Greasy Fleece Yield spread 5 years (2015-2019) were analysed. The overall least squares means for 1st six monthly, adult six monthly and adult annual GFY were 0.678±0.006 (1462) gms,0.484±.003(3995)and 0.972±.010(1694)kg, respectively. The period and sex had significant effect on all Greasy Fleece Yield.

Table 4: LSM of Wool production performance during 2015-16 to 2019-20 (farm flock)

Effect/trait	1st six monthly GFW	Adult six monthly GFW	Adult annual GFW
Overall (2015 to 2019)	0.678±0.006 (1462)	0.484±.003(3995)	0.972±.010(1694)
Year of shearing	**	**	**
2015-16	0.756±0.014(220)^{b}	0.511±0.006(782)^{b}	0.969±0.017(277)^{bc}
2016-17	0.716±0.011(381)^{b}	0.478±0.006(744)^{c}	0.979±0.017(276)^{bc}
2017-18	0.703±0.013(247)^{b}	0.479±0.005(908)^{c}	0.973±0.015(352)^{c}

Effect/trait	1st six monthly GFW	Adult six monthly GFW	Adult annual GFW
2018-19	0.611±0.012(288)[a]	0.440±0.005(1088)[a]	0.905±0.014(390)[a]
2019-20	0.602±0.012(326)[a]	0.514±0.005(1081)[b]	1.034±0.014(399)[b]
Sex	**	**	**
Male	0.719±0.008(657)	0.575±0.006 (747)	1.177±0.017(226)
Female	0.636±0.008(805)	0.394±0.002 (3856)	0.767±0.007(1468)
Season of shearing	**	**	-
Season1	0.624±0.008(698)	0.451±0.004 (2218)	-
Season2	0.731±0.008(764)	0.517±0.004 (2385)	-
Reg. of Age at shearing	*	**	**

Table 5: Wool quality (Mean±SE) parameters in farm flock2015-16 to 2019-20

Age group (N)	Fiber diameter (µm)	Medulation (%)			Staple length (cm)
		Hetro (%)	Hairy (%)	Total medulation (%)	
Adult					
Male	51.11±3.30 (10)	32.40±2.49 (10)	36.00±3.87 (10)	68.40±4.67 (10)	5.52±0.28 (10)
Female	43.85±1.80 (15)	36.84±2.10 (15)	30.64±3.29 (15)	67.49±3.11 (15)	5.59±0.17 (15)
Overall	46.75±1.81 (25)	35.07±1.64 (25)	32.79±2.51 (25)	67.85±2.58 (25)	5.56±0.15 (25)

3. Reproductive Performance

The reproductive performance of Malpura ewes during year 2010-2019 has been presented in Table 12. The improvement in tupping percentage, lambing percentage on available and tupped basis was recorded over the years. The tupping percentage was always more than 90 % in Malpura flock over the ten year. The lambing percentage was also maintained nearer to 85 percentage.

Table 6: Annual Reproductive performance (2010-11 to 2019-20)

Particulars	2010-11	2011-12	2012-13	2013-14	2014-15	2015-16	2016-17	2017-18	2018-19	2019-20
No. of ewes available for breeding	393	387	339	245	293	356	376	406	304	413
No. of ewes tupped		300	309	245	280	351	373	390	289	389
Tupping %	93.64	77.5	91.15	100	95.56	98.60	99.20	96.05	95.07	94.19
No. of ewes lambed		325	304	215	255	342	315	379	268	340
Lambing % on available basis	85.75	83.98	89.68	87.76	87.03	96.07	83.78	93.35	88.16	82.32
Lambing % on tupped basis	91.69	108.35	98.38	87.76	91.07	97.44	83.78	93.35	88.16	82.32

4. Survivability and Culling

The detail of annual survivability at different ages e.g. pre weaning (0-3 M), Post weaning, adult and annual are presented in Table 13. In general the survivability in all age groups was recorded to be above 90%. The detail culling at different ages e.g. pre weaning (0-3 M), Post weaning, adult and annual are presented in Table 14.

Table 7: Age-group wise survivability

Age group/ year	2009-10	2010-11	2011-12	2012-13	2013-14	2014-15	2015-16	2016-17	2017-18	2018-19	2019-20
0-3 Month	97.92	95.01	91.61	93.99	96.82	97.75	97.64	97.75	96.91	80.69	92.34
3-6 Month	95.67	97.41	91.88	87.05	95.86	97.81	97.43	98.96	98.75	99.97	97.01
6-12 Month							100	99.21	98.97	98.83	99.69
Adult	93.00	96.40	96.25	96.82	98.69	98.07	97.18	95.67	96.96	96.83	96.59
Overall	94.97	96.23	91.47	92.33	95.97	96.84	96.46	95.50	95.96	88.54	92.43

Table 8: Age-group wise culling

Age group/ year	2009-10	2010-11	2011-12	2012-13	2013-14	2014-15	2015-16	2016-17	2017-18	2018-19	2019-20
0-3 Month	00	00	0.65	0.21	0.00	0.23	0.00	0.17	0.30	0.00	0.00
3-6 Month	5.05	2.12	3.02	26.20	15.4	1.88	3.23	6.48	9.86	6.38	0.92
6-12 Month									2.33	3.21	0.62
Adult	4.17	10.12	11.25	21.10	10.08	8.87	12.50	17.90	12.69	9.36	6.35
Overall	7.74	5.01	8.71	19.29	12.5	6.64	8.49	14.90	7.98	5.96	2.87

5. Ram Distribution and Ewe Coverage

The superior breeding rams produced in the project were supplied to the Malpura unit farmers/State Animal Husbandry Departments /Govt. Agency/ Farmers/ NGO/Developmental agency for the genetic improvement of the animals in the farmers flock and a total of 631 superior breeding rams/ram lambs were supplied/sold to the Govt. Agency/Farmers/NGO/ Developmental agency during the reported period.

Table 9: Ram distribution

Year of distribution	No. of beneficiary farmers	No. of newly distributed &redistributed rams (1)	No. of rams sold for breeding purposes (2)	Total distributed/ re-distributed and sold rams (3=1+2)	Total no. of ewes covered by new+re-distributed rams
2010-11			129	129	-
2011-12			78	78	-
2012-13	6		51	51	-
2013-14	6		49	49	-
2014-15	31	33	-	33	1597
2015-16	9	12	-	12	574
2016-17	33	43	8	51	2026
2017-18	29	37	41	78	1503
2018-19	37	38	38	76	1916
2019-20	18	22	54	74	860
Total				631	

6. Genetic Parameter Estimates for Growth and Wool Traits

Data were analyzed regarding growth and wool traits of Malpura sheep, over a period 22 years from 1997 to 2018, recovered from the database of 5053 animals, maintained at ICAR- Central Sheep and Wool Research Institute (CSWRI), Avikanagar, Rajasthan. Heritability estimates from WOMBAT (animal model 1) method for body weights at birth, 3M, 6M, ADG1, ADG2 and GFW were 0.43±0.03, 0.13±0.03, 0.19±0.04, 0.12±0.03,0.08±0.03 and 0.29±0.04, respectively. It indicated that the genetic progress in these traits through pedigree selection would be faster than individual selection.Estimates of heritability from 1976-2019 for 6M was 0.136 ± 0.021.

Table 11: Genetic parameter estimates for growth and wool traits (1997-2018)

Trait	BWT	WWT	6WT	ADG1	ADG2	GFW
BWT	0.43±0.03	0.50±0.10	0.41±0.10	0.22±0.12	-0.009±0.16	0.38±0.09
WWT	0.38±0.02	0.13±0.03	0.92±0.04	0.95±0.01	0.41±0.21	0.28±0.13
6WT	0.34±0.02	0.78±0.01	0.19±0.04	0.90±0.05	79±0.10	0.39±0.11
ADG1	0.22±0.02	0.99±0.001	0.76±0.01	0.12±0.03	0.39±0.22	0.18±0.14
ADG2	0.02±0.02	-0.04±0.02	0.64±0.01	-0.04±0.02	0.08±0.03	0.34±0.16
GFW	0.23±0.02	0.31±0.02	0.40±0.02	0.28±0.02	0.24±0.02	0.29±0.04

Estimates of heritability (diagonal), genetic correlations (above diagonal) and phenotypic correlations (below diagonal) of different economically important traits of Malpura sheep

7. Genetic Gain

Individuals predicted breeding values were calculated for body weight at six months weights. The genetic trends were calculated by regression of average predicted breeding values for the traits (body weight at six months) versus the animal's birth year. Genetic gain for six month body weight was calculated from the inception (1975-2019) the genetic gain was 25.3g/year, it may be due to previously there was no genetic improvement programme was not initiated, during from the year 1997 genetic improvement programme was started. Genetic trend line is shown in figure as under and average breeding values as mean differences on yearly basis is presented in Table 10.

Table 13: Genetic trend for 6 month live weight (1975-2019)

Trait	Genetic gain per year
Six month body weight	25.3g

8. Inbreeding Coefficient

The data on 8124 individuals spanning over 45 years (1975-2019) of Malpura sheep were used for analysis of status of inbreeding in flock. Inbreeding in the flock was calculated by Wright's formula, and was computed from pedigrees traced in 7789 animals by path coefficient analysis using PEDIGREE VIEWER V6.5b computer package. The rate of inbreeding was calculated as $\Delta F_t = (F_t - F_{t-1})$ X100/(100-F_{t-1}),where F_t is the average inbreeding coefficient of individual born in period t. Average inbreeding over the period remain within limit i.e. 2.04%.The other parameter of inbreeding which is given in table .Overall inbreeding level is within the normal limit due to careful execution of matting plan by strictly following sire.

Table 14: Inbreeding level in the farm flock {(inception (1975) to 2019)}

Parameters	Value
Mean inbreeding	2.04%
Mean average relatedness	3.95%
Average generation interval (Lowest in 4.49 and highest in 5.76)	father to son and mother to son
Mean maximum generations	10.76
Mean complete generations	2.58
Mean equivalent generations	5.45
Realized effective population size	85.51
Effective number of founders (fe)	69
Effective number of ancestors (fa)	52
Effect of population bottlenecks (fe/fa)	1.32
No. of ancestors explaining 50% genetic variability	18

9. Impact Analysis

Improvement in body weights: The body weight at birth increases 39.17%, three month improved by 97.68 percent i.e. from 9.06 kg to 17.91 kg from inception to 2019. The body weight at six month improved by 88.82 percent i.e. from 13.77 kg to 26.00 kg from inception. The improvement (65.25 percent) in body weight at 12 month was also recorded as body weight improved from 18.40 kg to 29.91 kg.

Growth performance of Malpura sheep after 2010 is increasing significantly over the period after start, setting a new insight in the growth potential of native breed of Malpura sheep and possibility of its use as a mutton purpose sheep.

Survivability: Due to improvement in the body weights and proper management of flock the survivability of animals improved a lot over the years. In general, the overall survivability was more than 96 percent which is very much desirable for better remuneration to the farmers.

Lambing Percentage: The tupping percentage was always more than 90 % in Malpura flock over the ten year. The lambing percentage was also maintained nearer to 85 percentage since from last ten year.

Breeding rams: For the genetic improvement of the animals in the farmers flock and a total of 631 superior breeding rams/ram lambs were supplied/sold to the Govt. Agency/ MSSP Farmers/NGO/ Developmental agency during the reported period. Since 2014 onwards in Comparison of growth performance of progeny of distributed rams and farmer's rams, it shows that progeny of institute provided ram is always better than farmers ram.

10. Expected output

i. Genetically improved Malpura stock will be maintained at farm level

ii. Improved Malpura rams will be available for distribution in field.

11. Expected Benefits in Economic Terms

Malpura is one of the important sheep breeds among 8 breeds available in Rajasthan. The breed is well known for its mutton production. Ewes of this breed have good mothering ability and are having sufficient milk to sustain and accelerate the growth of their neonates. Lambs of this breed are quite suitable for fat lamb production.Malpura sheep is reared by small and marginal land holders who graze them on fallow land, crop residue and also take them on migration during period of scarcity. They are mainly reared for mutton purpose, as the earnings from their coarse wool are of little market value. Nowadays there is decrease in true to breed Malpura animals in the field so conservation of biodiversity has been a global issue and conservation of animal's genetic resources is a subject of great importance. It has been observed that Malpura breed is being replaced with Marwari or Marwari crosses in its home tact and this may threaten its survival. Keeping this in view to conserve this breed at institute farm near to its breeding tract for future use.

12. Risk Analysis

The traits of economic importance are highly influenced by the management and environmental factors such as feed, disease, natural calamity, etc. Performance of the flock depends on management. Sufficient technical manpower is required for effective monitoring of contractual services. In past few years there is paradigm shift of interest of sheep rearers from wool production to mutton production not only in semi-arid region but also in arid region of state as direct sale of early aged lambs fetches high prices; but due to sale of early ages of lamb the dressing percentage decreasing day by day. Possibilities need to be explored to rear the sheep in the field at least 6month age for higher mutton production.

13. Recommendation/ Future Programme

Malpura sheep is well known for its mutton production. Improving sheep for mutton production will have major emphasis on improving the indigenous Malpura sheep through selection. Breeding ewes and followers are envisaged to be kept for improvement and ex-situ conservation. Single trait selection using 6 month body weight was considered the bestselection criteria for this flock. Malpura sheep breeders is highly dependent upon the support provided by CSWRI Avikanagar in terms of supply of breeding rams, data collection, prophylaxis and treatment. The alternative model of community-based breeding program (CBBP) is proposed for self-sustainability and genetic improvement. With the formation of breeder's society and team of expert for selection of rams, sale and market linkage, the ram rearing center (RRC) will also be created which will be self-sustainable. CSWRI will continue to assist for selection decisions and real time data entry using cloud-based "Avimegh" platform. The early selection of lambs at weaning and rearing them at RRC till final selection at 12-month will provide unbiased selection of rams which will be given to farmers under CBBP for breeding. The surplus (80%) rams will be sold by RRC for profit. Several key measures have been suggested to make the Malpura CBBP self-sustainable and profitable entity.

19

Genomics of Heat Tolerance in Livestock with Heat Shock Proteins as Biomarkers

Samita Saini and Anju Kumari

Department of Animal Genetics and Breeding, Post Graduate Institute of Veterinary Education and Research (PGIVER), Rajasthan University of Veterinary and Animal Sciences, Bikaner Jaipur, Rajasthan

Livestock breeding for genetic improvement faces many challenges including feed conversion efficiency, fertility and adaptation to changing climate. These improvements have to be realized while maintaining or improving the nutritional properties of meat, milk and other animal products while giving emphasis to animal health and wellbeing also. Environmental stresses induced by climate change affect livestock productivity, reproductive efficiency and health, resulting in severe economic losses. Under the current climate change scenario, the viability of producing animals for human consumption is in jeopardy. In order to lessen the negative effects of heat stress in today's world, it is urgently necessary to develop methods that are centered on breeding animals with increased thermal tolerance and climate resilience. The breeding of animals having genetic potential to sustain under changing environment conditions can be achieved through the various molecular genetics tools.

Traditional Breeding Strategies

Breeds originating in hot climates show adaptation to heat stress compared with breeds that originated in temperate areas. Breeds from hot climates have lower respiration rates, body temperature, or sweating rates and better reproductive performance under heat stress. Locally adapted breeds have low level of production and the use of these breeds to improve heat tolerance of selected breeds will bestow an undesirable genetic condition in addition to the desired heat tolerance. When the production system provides adequate feeding, management, heat mitigation, and controlled parasite and pathogenic environment, selection for heat tolerance within highly productive breeds is beneficial than improving local breeds. In production systems that cannot provide mitigation for heat, adequate nutritional conditions or control of parasites and other pathogens, crossing of local and selected breeds and selection for productivity and monitoring of heat tolerance is the most excellent choice to improve productivity. Therefore the choice of breeding strategies to improve heat tolerance of livestock i.e., selecting within highly productive populations or crossbreeding/introgression will depend on the inputs of the production system

Heat Stress and Genetics

High temperatures and humidity levels cause reduced growth rates, reduced feed efficiency, decreased fertility, increased mortality, poorer fertility rates, increased abortion rates, and high culling. Heat tolerance is the ability of an animal to maintain production and reproduction levels under hot and humid conditions and is now a trait of economic relevance in animal production. Response of the animal to heat stress shows genetic variation due to which the production and reproduction of some animals is less affected by heat stress than of others and this response is heritable to progeny. The selective breeding for heat tolerance and rapid genetic gains may be possible using locally adapted breeds and crossbreeding. Gene introgression of desirable genes from one breed or subspecies into another can be done quickly by using genomic selection combined with advanced reproductive technologies. Therefore genetic selection might be a cost-effective tool to develop thermo tolerant animals but to be effective it requires a deep knowledge about the genetic basis of the animal's response to heat stress.

In order to understand the genetic component of thermotolerance and the genetic mechanisms behind the animal's response to heat, the following genetic analysis helps:

1. Association studies of polymorphisms at specific genes and genome-wide association analysis
2. Comparison of Genome between adapted and nonadapted breeds/species to harsh environments
3. Differential expression analyses.

Using these candidate genes, the heat stress response can be identified. Apart from the functional analysis of candidate genes for regulation of the heat stress response, families of genes present in association and differential expression studies can be identified. The most significant families are the heat shock proteins and DnaJs.

DnaJs proteins seem to be crucial partners of the heat shock protein-70 (Qiu et al., 2006) and they are important for protein translation, folding, unfolding, translocation, and degradation. In addition, genes from interleukin, chemokine, and fibroblast growth factor families are mostly involved in immunological and inflammatory processes resulting from the consequences for animals exposed to harsh environments.

Heat shock factor-1 is an evolutionarily conserved transcription factor that binds to the promoter regions of heat shock proteins to regulate their stress inducible synthesis in response to the environment. There are numerous candidate genes associated with regulation of the heat stress response like the slick hair gene in cattle and naked neck gene in poultry. The slick hair gene for smooth and short hair coat, located on chromosome BTA20, confers thermotolerance and is associated with an improved capacity for heat dissipation. Introgression of the slick hair gene from Senepol cattle and some lines of Holstein cattle produce animals with lower body temperatures and smaller declines in production under hot conditions. But it decreases the ability of animals to cope with cold temperature in climates that include both hot and cold periods.

The Naked neck gene Na gene is a single dominant autosomal gene that allows lessening of feathers in the neck region of bird, which allows better heat dissipation and shows better fitness under Heat Stress conditions. The Na genes diminish the plumage cover by 20% and 40% in heterozygous and homozygous Na, respectively. This gene is correlated with an increase in breast muscle and body weight, reduction in abdominal fat, and reduction in body temperature in broilers while in layer chickens it showed improved egg mass and egg characteristics. Na gene is responsible for the up-regulation of *HSP70* expression and has a positive impact on adaptation to heat stress.

Genomic Selection and Other Molecular Genetics Approaches

Genomic selection refers to the use of genome-wide genetic markers to predict the breeding value of selected animals and it relies on linkage disequilibrium between the markers and the polymorphisms that cause variation in traits. Among the various genetic markers used for genomic selection, SNPs are most commonly used markers in genomic selection. The prediction equation of breeding values for a trait from SNP genotypes must be estimated from the reference population which has been measured for the traits and genotyped for the SNPs. This prediction equation can be used to predict breeding values for selection candidates based on their genotypes alone. The animals are ranked on the basis of estimated breeding values, and the best ones are selected to breed the next generation. Genomic selection has advantage over traditional methods of selection based on pedigree and phenotype alone, to facilitate the selection of animals done accurately early in life, based on their genomic insight, for traits that are difficult or expensive to measure such as fertility disease resistance, methane emission and feed conversion. Genomic selection can also be done for development of animals showing resistance to heat stress.

The recent improvement in molecular biology has led to the development of various biotechnological tools like next-generation sequencing (NGS) genome-wide association studies (GWAS) , whole transcriptome analysis and microarray technology which can be used to quantify the molecular mechanisms which govern the climate resilience capacity of animals. GWAS is the most preferred option among these for researchers as this approach mutually takes into account all the genotype, phenotype, and pedigree information of farm animals. Selection signatures also help to differentiate functionally significant regions in the genome which can be used to detect prospective loci and candidate genes that have undergone positive selection. These identified biomarkers can be integrated into the existing breeding policies using genomic selection to develop climate-resilient livestock.

Heat Shock Proteins as Biomarkers

Biomarkers are molecules that show signs of an alteration in protein or gene expression. Heat shock protein, whose expression is noticeably elevated during heat shock, is a critical biomarker. Ferruccio Ritossa, an Italian geneticist, is credited for publishing the first study on heat shock protein. He did it at the Genetics Institute in Pavia, where he was working on Drosophila (Ritossa, 1996). The identification of genes containing thermo-tolerance in indigenous breeds can greatly aid in the genetic development of animals for the creation of adaptive and productive types. The heat stress (HS) response is a highly conserved cascade of protein activation and altered gene expression that occurs in response to a variety of stressors (Sonna et al., 2002; Collier et al., 2006). The HSP27, HSP60, HSP70, and HSP90 families are the most common families and their expression has been detected in gametes and early-stage embryos (Wilkerson and Sarge, 2009).

Heat-induced Cellular and Molecular Responses

Thermal stress causes multiplicity of abnormalities in the functioning of the cells, like changes to biological molecules, disruptions in cell processes, metabolic reactions, oxidative cell damage, and activation of the apoptosis and necrosis pathways. Depending on the timing and effectiveness these abnormalities can consequence in cell survival, acclimation, or death. HSPs play role in cytoprotection, as indicated by prevention of hyperthermia, circulatory shock, and cerebral ischemia during heat stroke. Due to an excess of free radicals and reactive oxygen species (ROS), oxidative stress damages the antioxidant defence. There is an increase in ROS generation, particularly the superoxide anion (O2) and increased levels of lipid peroxidation byproducts, malondialdehyde (MDA) and thiobarbituric acid-reactive species (TBARS), in broilers, buffaloes, and dairy cows. Superoxide dismutase (SOD), catalase (CAT), and glutathione peroxidase (GPX) antioxidant enzyme activity were found to increase in HS in animals.

The effects of heat stress can be seen in the modification of multiple molecular functions, including DNA synthesis, replication, and repair; cellular division; the functions of nuclear enzymes and DNA polymerases; and the fluidity and stability of cellular membranes, as well as the inhibition of the function of receptors and transmembrane transport proteins. Heat stress can also affect nuclear enzymes and DNA polymerases. The cell shape, the mitotic apparatus, and the intracytoplasmic membranes such as endoplasmic reticulum and lysosomes are some of the aspects of the cytoskeleton that change as a result of the induction of various variants in the organisation of the cytoskeleton by heat stress.

When animals are subjected to heat stress, a wide variety of cellular and molecular reactions are triggered. Because of advancements in molecular biotechnologies, there are now new options to define gene expression and determine essential cellular responses to heat stress. Alterations in the expression patterns of particular genes that are essential for thermo-tolerance at the cellular level in animals and are having a function of cellular adaptation are thought to be possible biomarkers for identifying the stress adaptation mechanisms that are at play. When cells are subjected to heat stress, several genes, including those encoding heat shock proteins (HSP), apoptotic genes, and other cytokines and toll-like receptors, are activated. Other genes, such as those encoding superoxide dismutase (SOD), nitric oxide synthase (NOS), prolactin receptor (PRLR), and thyroid hormone receptor (THR), were discovered to be related with thermotolerance in ruminant livestock.

Because of the adaptive mechanism to combat both heat stress and nutritional stress, there is a higher expression of HSP70 messenger RNA (mRNA) in the adrenal glands of multiple stressor groups. This is owing to the fact that HSP70 is a heat shock protein. The considerably higher expression of adrenal HSP70 mostly in multiple stressed animals as compared with just heat stressed animals could be attributed to additional nutritional stress, as well as hyperactivity of the adrenal cortex to produce more cortisol. This comparison was made by comparing the multiple stressed animals to animals that were only exposed to heat stress.

Components of Heat Shock Response and Their Importance in Livestock Adaptation

Heat shock transcription factors Families (HSF) that are controlled by inducible expression of HSF genes manipulate the cellular response to thermal stress in mammalian animals at the transcriptional level. Diverse livestock species have different HSF isoforms. The unfolded protein's hydrophobic domain, which experiences trimerization in the nucleus and hyper phosphorylation during HS, first activates the isoform HSF1. After activation, the HS elements (HSEs) are bound by the HSFs, which increase the transcription of HSP mRNA. Large eukaryotes have been shown to have HSF1 to HSF4 out of which HSF1 has received the most interest and is directly related to livestock's ability to tolerate heat. Only chickens contain the HSF3. While HSF2 is triggered in response to cellular stress other than thermal stress, HSF1 and HSF3 are active during HS. The HSF1 and stimulation of HSP70 gene expression are mainly connected.

HS is the result of temperatures that are higher than 25 degrees Celsius and is the body's reaction to stimuli that disrupt homeostasis, which for the majority of farm animals is generally between 4 and 25 degrees Celsius. Thermoregulation is the process that allows livestock species to maintain a typically consistent body temperature by balancing the systems in their bodies that create heat and remove heat in order to keep their internal temperature relatively constant. Animals will suffer from hyperventilation syndrome (HS) if the energy balance is disrupted and the temperature humidity index (THI) climbs above 72. When farm animals are subjected to environmental stress, they selectively manufacture a number of proteins, including the gene for slick hair (ATP1B2), heat shock proteins, and other proteins in the heat shock protein family (HSPs). At the level of the cell, the inducible version of the heat shock protein HSP70 known as HSP70i has been suggested as a potential predictor of thermotolerance in livestock animals.

Types of HSPs

On the basis of their molecular weight and the biological roles they perform, HSPs are divided into a number of different families, including HSP 110, HSP 100, HSP 90, HSP 70, HSP 60, HSP 40, and HSP 10. In livestock species, the development of thermo tolerance is primarily correlated with HSP70 and HSP90. Cattle, buffalo, sheep, goats, and broilers all had higher levels of HSP70 and HSP90. The most abundant and temperature sensitive HSP is HSP70 (HSP70-1 and HSP70-2). Goats show an elevated expression of many heat shock proteins (HSPs) when subjected to hypothermic stress. Some of these heat shock proteins are HSP32, HSP40, HSP60, HSP70, HSP90, and HSP110. It has been observed that the blood levels of HSP-70 increased in poultry when they were exposed to HS, and it has also been reported that the decrease in spermatogenic cells in the testis of the stressed group is attributable to HSP-70 suppression.

HSP Expression Mechanism

HSF1 monomer which is bound to HSP in the cytoplasm during unstressed conditions, dissociates and binds with other HSF monomers for trimerization before nuclear translocation during heat stress. When homotrimeric HSF binds on the HS element (HSE) in the nucleus and hyper-phosphorylation, the HS target gene transcription is activated, resulting in increased expression of HSP mRNA. HSP biosynthesis and chaperonic function are energy-seeking processes. HSP70 is involved in three distinct but related processes in the majority of living creatures. These processes are the prevention of protein aggregation, the facilitation of the folding of misfolded proteins back into their natural state, and the solubilization and refolding of aggregated proteins. There are at least four members of the HSF gene family that are expressed in plants and vertebrates alike, and there are three HSFs that have been found in humans (HSF1, HSF2, and HSF4). It was discovered that the nucleotide sequences of caprine HSP70 genes appeared comparable to those of cow and pig HSP70 genes, followed by those of horse, camel, and human HSP70 genes. This similarity indicates a tight evolutionary link among the HSP70 family of genes. HSPs, unlike other proteins, do not denature under stress because they have enhanced secondary protein structure, better hydrophobic packing, stronger hydrogen bonds, and helix dipole stabilization.

Heat Stress Biomarker HSP70 in Animals

HSP70 has been classified into constitutive and inducible forms (HSP70i), both of which relieve stress through molecular chaperon activity in the cytoplasm. Numerous in vitro and in vivo neurodegenerative animal models have also shown the role of HSP70 in neuroprotection. Potentially varying thermo-tolerance among species could account for observed differences in HSP70 expression between organisms. HSP70 not only enhances overall protein integrity, but it also inhibits cellular apoptosis, essentially protecting cells from stress. It was found that HSP70 might be used as a biomarker for heat stress and a reduced immunological state in buffalo heifers (Patir and Upadhyay, 2010). HSP70 plays a role in reducing heat stress while maintaining cellular homeostasis. HSP70 plasma level is a robust biomarker of longterm stress in the case of repeated stresses but is not a reliable predictor of single stressor severity due to its strong correlation with environmental temperature but not with animal body temperature (Gaughan et al., 2013).

Cytoprotection by HSP70 in Animals Under Heat Stress

HSP70, like other HSP proteins, is a highly conserved protein that is activated by various stresses, including thermal stress. HSP70 inhibits cytotoxic protein complexes during heat stress by interacting with unfolded/misfolded (denatured) proteins, restoring cellular protein homeostasis. HSP70-mediated cytoprotection has been observed in many animal organs, including the intestine, kidney, and embryo as increased expression of HSP70 in animals suffering from heat shock, circulatory shock, and cerebral ischemia. High levels of heat shock proteins can also be induced by infection, inflammation, exercise, cell exposure to harmful materials such as UV light, starvation, hypoxia, nitrogen deficiency (in plants), or water deprivation. As a result, heat shock proteins are also known as stress proteins, and their up regulation is sometimes referred to more broadly as part of the "stress response" or "heat shock response". Several heat shock proteins act as chaperones for other proteins within the cell.

Heat-shock proteins can also be found in non-stressful conditions, simply "monitoring" the proteins in the cell. Their role as "monitors" includes transporting old proteins to the cell's "recycling bin" (proteasome) and assisting newly synthesised proteins in folding properly. These activities are part of the "heat stress response," a cell's own repair system. HSPs appear to play an important role in the cardiovascular system also. HSPs play a role in immunity due to their ability to bind not only whole proteins, but also peptides, which determines their function in immunity.

Other Applications of HSP

Cancer Vaccines and Anticancer therapeutics- HSPs can be used as immunologic adjuvants (DAMPS) to boost a vaccine's response and may be useful in improving the efficacy of cancer vaccines. Intracellular heat shock proteins are highly expressed in cancerous cells due to presence of mutated and over-expressed oncogenes. HSP levels in the blood are excellent biomarkers for determining the stage and aggressiveness of certain cancers. It has been demonstrated that the expression of HSP27 and HSP70 can affect the way in which tumour cells react to traditional cancer treatment. Greater expression of HSP27 leads to a favourable outcome for chemotherapy in patients with breast cancer and leukaemia, whereas increased expression of HSP70 leads to a favourable outcome for chemotherapy in patients with osteosarcomas.

Autoimmunity treatment-HSPs, which act as DAMPs, can extracellularly promote autoimmune reactions that lead to diseases such as rheumatoid arthritis and systemic lupus erythematosus. Administering some HSPs to patients can induce immune tolerance and treat autoimmune diseases. HSPs particularly hsp60 and hsp70 are being studied in clinical trials for the treatment of rheumatoid arthritis and type I diabetes.

Factors Affecting Induction of HSP70

Heat shock is not the only factor that contributes to the induction of HSP70; other factors also play a role in this process. Protein kinase C stimulators, glucose analogues, heavy metals, exposure to amino acid analogues, calcium raising mediators, ischemia, sodium arsenite, microbial disease, nitric oxide, hormones, and antibiotics are some of the additional stimuli that can cause a rise in the amount of HSP.

Potential of HSP70 as a Tool for Selection of Thermo-tolerant Animals

There is a wealth of research that points to concrete evidence of enhanced expression of HSP70 in response to heat stress in a variety of animals. This serves as a protective mechanism against the negative consequences of stress. The HSP70 family is a possible candidate genetic marker that can be used in the selective breeding of animals that can withstand high temperatures. It has been shown that a single nucleotide polymorphism (SNP) in the promoter region of the cattle HSP70-1 gene can alter gene expression. This change in gene expression has been associated with increased thermotolerance and milk output in Frieswal cows (Flynn et al., 1991). There is a correlation between the 3' -UTR region of HSP70 in swine and both mRNA stability and the response to stress (Schwerin et al., 2001). It is impossible to overestimate the importance of the SNP located in the promoter region of the HSP70 gene as a reference for selecting dairy cattle for their thermotolerance (Deb et al., 2015). A more recent study discovered a correlation between a single nucleotide polymorphism (SNP) in the 5'-untranslated region (UTR) of the HSP70 gene in cattle and the number of services provided per conception in Pasundan cows (Said and Putra, 2018). There is a correlation between different mutations have been identified in HSP genes and different reproductive characteristics in bovines. It is now possible, as a result of recent advances in biotechnology in molecular genetics, marker assisted selection, and transgenesis, to deconstruct genomic variations in animals and use them for selective breeding in order to produce subsequent generations that are better adapted to their environments.

The genetic development of future generations in terms of thermotolerance is a more sustainable and rational strategy than increasing animal production management. This is because genetic improvement may be passed on from generation to generation. As a consequence of this, it has been suggested that various HSP70 variations in flanking and UTR regions be utilised as genetic markers for the identification of more adaptable animals. This will unquestionably contribute to an improvement in both performance improvement and animal welfare.

Conclusion

When the ambient temperature rises, gene networks inside and across cells and tissues react by sending intracellular and extracellular signals that coordinate the metabolic processes of individual cells as well as the animal as a whole. When animals begin to store heat and rapidly accelerate evaporative heat loss processes at skin surface temperatures that are higher than 35 degrees Celsius, these systems begin to be activated. Examples of changes in gene expression include the activation of HSF1, an increase in HSP expression, a decrease in the expression and synthesis of other proteins, an increase in the metabolism of glucose and amino acids while a decrease in the metabolism of fatty acids, the activation of the stress response in the endocrine system, and the activation of the immune system via HSP extracellular secretion. If the stress continues, the altered gene expression will lead to an altered physiological state known as acclimatisation. Acclimatization is mostly controlled by modified gene expression in response to endocrine signals. Acclimation can only occur if the stress continues. Because of the wide range of evaporative heat loss between animals and the central role that HSF1 plays in coordinating thermal tolerance, it appears that there may be room for improvement in cattle thermo-tolerance if the genes that control HSF1 expression and those that regulate evaporative heat loss are manipulated.

Livestock rely heavily on the cellular response as a means of overcoming the effects of heat stress. This is the mechanism by which an organism can maintain its health when under pressure. The ultimate goal of this route is the production and discharge of HSPs. There are three parts to the heat shock response: HSF, HSE, and HSP. Of the several HSPs found in farm animals, HSP70, HSP90, and HSP27 have been the subject of the most research. Increased animal output and productivity will benefit farmers and the country's economy if scientists can better grasp the biological mechanisms of different heat shock proteins (HSPs) under heat stress. When comparing HSPs in farm animals, HSP70 was determined to be the most accurate biological marker for heat stress.

Because of their role as molecular chaperons in DNA repair, apoptosis, signal transduction, and protein homeostasis, animals that overexpress HSP70 are able to endure high temperatures for longer periods of time. HSP70 in the extracellular fluids has a role in immunomodulation by promoting the creation of neutrophils, macrophages, and anti-inflammatory cytokines like IL-10 to alert immune cells against invading pathogens. Calf weaning weights, pregnancy rates, and fertility, in especially in dairy cattle, have all been linked to variations in the HSP70 5'-UTR and promoter region. Specifically, the highly variable areas of HSP70 genes that are linked to thermo-tolerance and performance attributes could be used as selection markers. Selective breeding to increase thermo-tolerance in future generations is considered a more long-term and enlightened technique than bettering animal production management. A member of the heat shock protein 70 (HSP70) family may serve as a useful marker for improving cattle productivity around the world by identifying animals that can withstand extreme temperatures and respond effectively to disease.

References

Collier RJ, Dahl GE, VanBaale MJ 2006. Major advances associated with environmental effects on dairy cattle. J Dairy Sci;89:1244e53.

Deb R, Sajjanar B, Pavani KC (2015) Bovine heat shock protein 70 and its application in cellular thermo tolerance. J Vet Sci Technol 6(6): 1000e121.

Flynn, Gregory C., et al. (1991): "Peptide-binding specificity of the molecular chaperone BiP." Nature 353.6346 726-730.

Gaughan JB, Bonner S, Loxton I, Mader T. 2013 Effects of chronic heat stress on plasma concentration of secreted heat shock protein 70 in growing feedlot cattle. J Anim Sci;91:120e9. https://doi.org/10.2527/jas.2012-5294.

Hayes, B.J. et al. (2009) A validated genome wide association study to breed cattle adapted to an environment altered by climate change. PLoS ONE 4, e6676

María J Carabaño, Manuel Ramón, Alberto Menéndez-Buxadera, Antonio Molina, Clara Díaz 2019 Selecting for heat tolerance . Animal Frontiers, Volume 9, Issue 1, January 2019, Pages 62–68

Patir H and Upadhyay RC. 2010; Purification, characterization and expression kinetics of heat shock protein 70 from Bubalus bubalis. Res Vet Sci 88:258-62.

Ritossa, Ferruccio (1996) "Discovery of the heat shock response." Cell stress & chaperones 1.2 97.

Said S, Putra WPB. 2018 Novel single nucleotide polymorphisms (SNPs) in the 5'UTR of bovine heat shock protein 70 (bHsp702009) gene and its association with service per conception (S/C) of Pasundan cattle. Biodiversitas;19:1622e5. https:// doi.org/10.13057/ biodiv/d190504.

Schwerin M, Czernek-Schafer D, Goldammer T, Kata SR, Womack JE (2003) Application of disease-associated differentially expressed gees, mining for functional candidate genes for mastitis resistance in cattle. Genet Sel Evol 35: S19-S34.

Schwerin M, Maak S, Kalbe C, Fuerbass R (2001) Functional promoter variants of highly conserved inducible hsp70 genes significantly affect stress response. Biochim Biophys Acta 1522(2): 108-111.

Sonna LA, Gaffin SL, Pratt RE, Cullivam ML, Angel KG, Lilly CM (2002) Effects of acute heat shock on gene expression by human peripheral blood mononuclear cells. J Appl Physiol 92:2208– 2220

Wilkerson DC, Sarge KD. RNA polymerase II interacts with the Hspa1b promoter in mouse epididymal spermatozoa. Reproduction;137:923e9.

20

Genomic Selection A Paradigm Shift in Revolutionizing Animal Breeding

Vaishali Khare, Ankur Khare, Shreekant Joshi, Satyendra Singh Tomar and Vikas Mahajan

Department of Animal Genetics & Breeding, College of Veterinary Science and Animal Husbandry, NDVSU, Jabalpur

Introduction

Estimated breeding values (EBVs), often estimated for sires, are used in traditional animal breeding (TAB) programmes to make selections based on an animal's performance and that of its relatives. Using, typically, the best linear unbiased projections, EBVs are calculated using attributes measured in own-performance, pedigree, sib, and/or progeny evaluation schemes in specialised testing stations and/or selected farm conditions (BLUP; Henderson, 1975). One of the main time-consuming challenges for phenotypic evaluation to enable a meaningful estimation of breeding values is generation intervals of 4 to 5 years in beef cattle, more than 5 years in dairy cattle, and 2 to 2.5 years in pig breeding schemes (Schefers and Weigel, 2012). Modern breeding techniques like marker-assisted selection (MAS) and, more recently, genomic selection (GS) are based on the use of genetic DNA markers to aid in selection and breeding (Williams, 2005). The fundamental reason why MAS, that only further uses a few markers as a selection tool, hasn't made as much progress as was initially anticipated is that finding trustworthy markers is challenging, particularly when dealing with complex traits. Only relatively few consistent markers could be identified, despite successful identification of many quantitative trait loci (QTL; Hu and Reecy, 2007).

In 2001, Meuwissen *et.al.,* proposed a novel approach " Genomic Selection" where the breeding value could be estimated from markers spanning the entire genome. With this approach, genetic effects are estimated for each marker and then summed up to predict the overall breeding value of any animal. This estimation of marker effects is carried out within a reference population, i.e. a large group of individuals with both phenotypes and marker genotypes information. These effects are then applied to candidates for selection with marker genotype information, but without known phenotypes. High density markers covering the entire genome are employed simultaneously for individual genetic evaluation via genomic estimated breeding values (GEBVs) in the marker-assisted selection technique known as genomic selection (GS). By selecting individuals early in life, GS can improve selection accuracy, reduce the generation gap, and hasten genetic development. High density whole genome SNP (single nucleotide polymorphism) chips for livestock are now available, and GS is changing the way that traditional animal breeding and selection processes are done. GS is quickly becoming the most important method of genetic assessment for bull selection in dairy cattle in many countries, and it is likely that it will soon entirely replace the conventional genetic evaluation system in dairy cattle. In recent years, GS has emerged as a key research area in the breeding of animals, plants, and aquatic organisms, and many fascinating findings have been revealed as a result of this work.

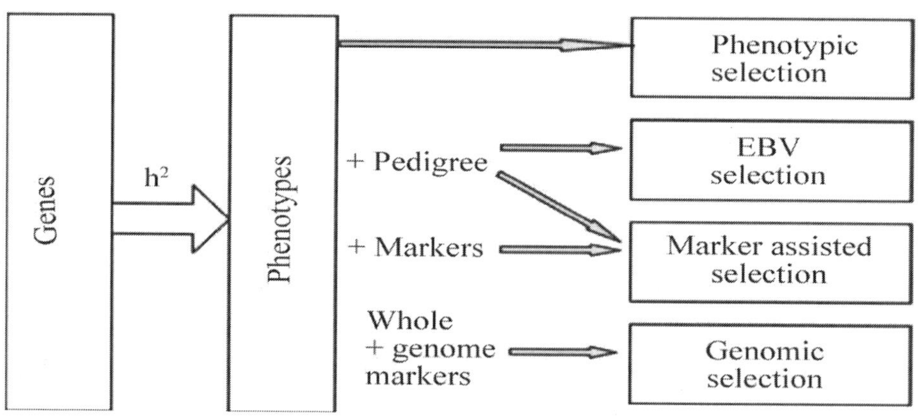

Fig. 1: Schematic diagram of selection methods in animal breeding (Zhang et. al., 2011)

Traditional Animal Breeding

In TAB, breeders set a breeding goal based upon desirable production traits specific for a given breed viz milk production traits in dairy cattle and meat production in beef cattle. In each breeding generation, sires and dams are phenotypically scored (using either quantitative or qualitative criteria), with the scores for each trait generating a BV that is used to rank the animals on genetic merit. Sires and dams with the highest BVs or in other words with most desirable physical characteristicsare selected for breeding. Offspring from these matings are evaluated further according to the set breeding goal and those with the highest BV scores are used as breeding stock for the next generation.

However, the best phenotypic assessments cannot ensure with certainty that mating two high BV animals will result in high BV offspring. Therefore this process of assessing individuals is limited by many factors that includes:

- Limitations in understanding of most traits at the biological level due to complex inheritance (rather than simple Mendelian inheritance) and number of traits that can be assessed in a breeding goal
- Decreased selection efficiency for traits that are difficult to measure (e.g., meat quality, internal parasite resistance) or traits with low heritability (e.g. reproduction)
- Restriction to traits that can be reproducibly and inexpensively measured in a large number of animals
- Genetic progress is slow, and a high selection pressure on key production traits may have a deleterious effect on other important traits that are not readily scored

Marker-Assisted Selection

Marker assisted selection or marker aided selection (MAS) is an indirect selection process where a trait of interest is selected based on a marker (morphological, biochemical or DNA/RNA variation) linked to a trait of interest (e.g. productivity, disease resistance, abiotic stress tolerance, and quality), rather than on the trait itself. Because of the above, MAS has the potential to significantly speed up the process of genetic improvement by increasing the accuracy of selection while simultaneously decreasing the amount of time required. When screening for the trait's phenotypic is challenging or when the trait is manifested late in life, MAS becomes very useful. The ability to analyse a large number of individuals in a reasonable amount of time and at a reasonable cost is essential for a successful MAS strategy, as is the availability of a genetic map with a sufficient number of regularly spaced markers, tight linkage between the gene or QTL of interest and adjacent markers, and adequate recombination between the markers and the remainder of the genome. In addition, MAS is most effective when integrated into conventional breeding programmes that already keep track of performance and pedigree information.

Genomic Selection

Genomic selection (GS) is a refined form of MAS made possible by genomics capabilities that allow for the simultaneous screening of tens of thousands of genetic markers in the form of SNPs. In this method, genomic and phenotypic information from a sample of the population is used to create a "training set," or set of data, from which statistical associations between markers and phenotypes can be derived, and equations for the prediction of breeding values can be created. These equations are then used to predict breeding values for individuals for whom no phenotypic information is available, or the "validation set" (Meuwissen et al., 2001). GS does not require *a priori* phenotypic information for the validation set and thus can increase the genetic gain by accelerating the breeding cycle.

Difference Between MAS and GS

The fundamental difference between MAS and GS impacting the effectiveness of these two selection tools is 'number of QTL' used. MAS is limited in its ability to predict breeding values as it concentrates on a small number of QTLs that are tagged by markers with well-defined associationsIn contrary, GS makes use of a large number of markers that are spread out across the entirety of the genome. This ensures that each and every QTL is linked in LD to at least one SNP marker.

Principle of Genomic Selection

A commercial or bespoke SNP array is used to determine the genotype of a reference population of animals after they have been assessed for important production traits. The values of the variable (x), which reflect the genotypes, consist of 0, 1, 2 (homozygous, heterozygous, or alternate homozygous). For each animal, a gEBV is calculated by creating a prediction equation that takes into account the genotypes of all the markers and the effects of those genotypes. This equation for making predictions can be used to a group of animals that have not been phenotyped, breeding values can be computed, and the most desirable individuals can be chosen for breeding (Goddard and Hayes, 2007).

In contrast to conventional MAS, which predicts breeding values by using only those few markers that are thought to have large effects on the underlying trait of interest, GS simultaneously uses high density (HD) markers covering the whole genome to predict the breeding values of every genotyped individual. Traditional MAS can only predict breeding values by using those few markers (Goddard and Hayes.2007 The term "genome-estimated breeding value" is used to describe a breeding value anticipated in this way (GEBV). GS operates under the hypothesis that all genes or QTLs (quantitative trait loci) involved in the characteristic of interest are in LD (linkage disequilibrium) with at least one of the markers. If the marker density is high enough, all genetic variance should be able to be tracked, theoretically.

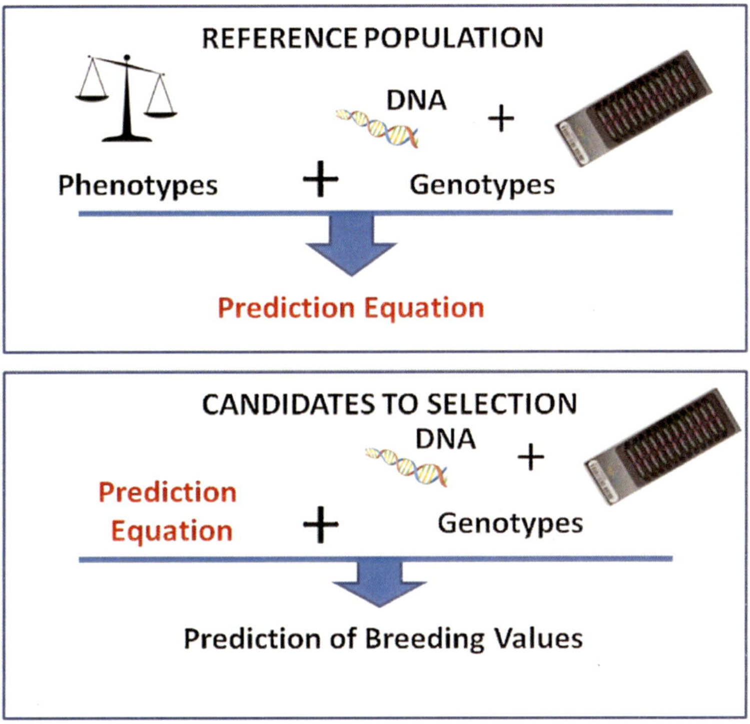

Fig. 2: Principles of genomic selection. Top: a prediction equation is obtained from a reference population with phenotypes and genotypes; bottom: this prediction equation is used on candidates with genotypic information only. (Boichard et al., 2016)

Direct and Indirect GS

GS approaches can be broken down into two categories, indirect and direct, depending on the type of statistical model employed. Indirect approaches first estimate marker effects in a reference population, and then use this information to determine the GEBVs of genotyped individuals in the candidate population. The GEBVs are directly calculated via direct methods, which employ mixed model equations (MME) in the same way that the conventional BLUP approach does. An HD marker-derived connection matrix is built utilizing the phenotype data of the reference population and the genotype data of all the markers from the reference and candidate populations.

Figure 3 shows a diagram of the indirect GS method and the direct GS method.

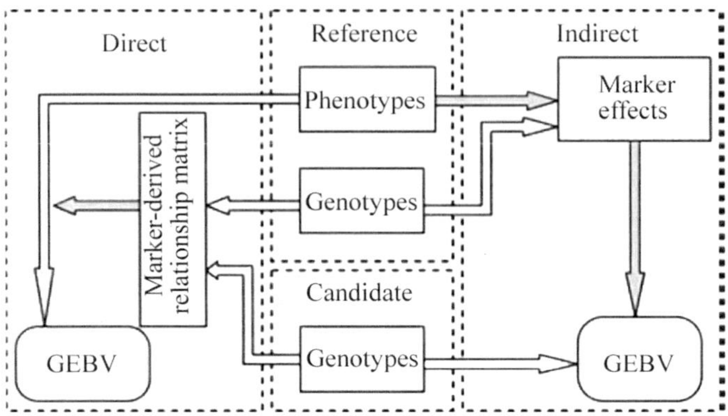

Fig. 3: Schematic illustration of the indirect and direct GS methods. (Zhang et.al., 2011)

Factors Affecting Genomic Selection Efficiency

Whatever the domestic species targeted for selection, the yearly genetic gain depends on four parameters: Genetic variability of the trait, selection intensity, evaluation accuracy and generation interval. Among these except for first one, the three can be modified by genomic selection. The main advantage of genomic selection is that candidates can be evaluated and therefore selected without their own phenotypic information, nor on their progeny. Therefore, selection can be applied very early, just after birth or even on embryos. Therefore, in a number of species and production systems, genomic selection may rely on a reduced generation interval. When the genotyping cost is low, a large number of candidates can be screened, and selection intensity can be increased. This large-scale screening also allows a better use of the available genetic resources. The evaluation can be carried out for any trait recorded in the reference population, which is of particular interest when the trait is difficult or impossible to record on the candidate itself (sex-limited traits, meat quality traits, disease resistance). The third determinant of selection efficiency, genomic evaluation accuracy (r) depends, for a given trait (1) on the accuracy of SNP effect estimation and (2) on the linkage disequilibrium between SNP and causal variants. The first parameter depends on the size (N) of the reference population and on the heritability of the trait (h^2). The second parameter depends on the structure of the genome and the genetic architecture of the trait. A critical parameter is the number of independent segments segregating in the population. This number q is a function of the length of the genome (length = L, in Morgans) and of the effective size of the population Ne. When Ne is smaller, the conserved segments are longer and fewer markers are needed to trace them. In practice, a formula is commonly used to evaluate this accuracy (Daetwyler et al 2010)

$r^2 = N h^2/(N h^2 + q)$, with q a function of Ne and L.

Current Status of Genomic Selection

When the conventional progeny test system is replaced with a GS breeding programme, a simulation study found that dairy cattle breeding organisations might save as much as 92% on breeding costs (Schaeffer 2006). The considerable decrease in generation interval and the improvement in selection accuracy for bull dams are the primary causes of this cost-savings (Goddard and Hayes,2007). In recent years, there has been a global uptick in the adoption of GS in animal breeding programmes due to this intriguing promise, which is undoubtedly appealing to breeders and researchers.

Genomic Selection in Dairy Cattle

GS has been utilised in national and international dairy cattle breeding programmes in numerous nations, including the United States and Canada (Hayes et al., 2009), Australia (Moser et al., 2009, Hayes et al., 2009), Norway (Luan et al., 2009), New Zealand (Harris et al., 2009), The Netherlands (de Roos et al., 2009), Denmark (Lund et al 2009),Ireland (Berry et al 2009) and Germany (Habier et al 2010). All of the features that are being targeted for improvement through traditional breeding methods are included in those that are currently being examined with genomic information. Individuals that had their genomes sequenced varied from bulls to milking cows to heifers.

Genomic Selection in Other Livestock Species

Despite the fact that HD SNP panels for swine (Illumina PorcineSNP60), sheep (Illumina OvineSNP50), horse (Illumina Equine SNP50), and chicken (Illumina iSelect 18K Custom genotype) are currently available, there has been very little reported deployment of genetic screening in these species.

In the case of sheep, a theoretical calculation revealed that GS had the potential to increase the overall response for a terminal sire index by approximately 30% and for a fine wool merino index by approximately 40%. Both of these increases were in comparison to the traditional selection methods that are commonly used. According to research published in 2010 by Daetwyler et al., the accuracies of GEBV for wool traits in Merino sheep ranged from 0.15 to 0.79, while the accuracies for meat traits in Merino sheep ranged from -0.07 to 0.57. Both populations were comprised of 7180 sheep.

When compared to a pedigree index, Gonzalez-Recio et al. (2009) discovered that GS could boost the accuracy of selection for food conversion rates in broilers by up to almost four times. It has been demonstrated that applying GS in layers can boost the accuracy of selection by up to 88 percent when done again at a later age, as well as by up to double when applied at a younger age (Wolc et at., 2011).

Traditional breeding plans for pigs have mostly focused on the performance characteristics of the sire lines, however genomic selection in pig breeding offers the possibility of improving maternal attributes. GS has been shown in a simulated research to have the ability to boost genetic gain for maternal qualities in pigs, which would be beneficial (Lillehammer et al., 2011). An economic feasibility analysis of GS implementation in male lines found that the quality of phenotypic selection might be increased while using the same (or a lower) budget than conventional methods (Tribout et al., 2012). Genomic selection could potentially play a role in aiding the increase of genetic gain in developing nations, where both native and exotic genetic lines could be exploited for several generations of backcrossing (Akanno et al., 2013). The utilisation of genomic information will, in addition, be of assistance in such schemes for controlling inbreeding and preventing the potential loss of genetic variation.

Impact of Genomic Selection on Traditional Breeding Systems

Effect on current breeding schemes: In GS, the link between individuals can be inferred using whole-genome markers, as opposed to the pedigree-based methods used in traditional genetic evaluation systems. There could be re-percussions for animal breeding as a result of this. Breeding herds, multiplier herds, and production herds are the standard components of a swine or poultry breeding system. Since it is not possible to utilize pedigrees to establish the links between animals in the production and breeding herds, the performance records of hybrid animals in production herds are rarely used to evaluate the animals in breeding herds. However, with the use of HD markers and data collected from as many herds as feasible, it is simple to establish a link between animals from various strata (Ibanz et.al., 2009). When it comes to dairy cattle, the progeny testing breeding system requires a waiting period of at least six years before a tested bull can be utilized for breeding purposes. A plan to breed animals in this manner would not be financially viable. It is possible to conduct early stage selection of young bulls based on their GEBVs using GS, which results in a reduced generation interval and lower breeding costs (Konig et.al., 2009).

Joint Breeding of Multiple Populations and Multiple Breeds: In order to examine the reliability of genomic prediction across multiple populations, de Roos et al. 2009 created simulated reference populations made up of two cattle populations that diverged 6, 30, and 300 generations ago. These populations were compared to one another. Toosi et al., 2010 and Ibanz et al., 2009 studied genetic structure in admixed and crossed populations by modelling purebred as well as F1, F2, 3- and 4-way crossings. Their findings suggested that GS may be used to a number of different populations as well as different breeds. GS can be carried out in crossbred populations, and models that fit the breed-specific effects of SNP alleles might not even be necessary, especially when dealing with HD markers (Ibanz et al., 2009). A multiple breeds GS analysis was carried out by Kizilkaya et al., 2010 using simulated phenotypes and genuine 50K SNP genotypic data. The accuracy of GEBVs was studied by Hayes et al. 2009 in Holstein and Jersey dairy cattle, with the reference population consisting of either a single breed or a multi-breed combination. They found that a reference population consisting of more than one breed performed significantly better than one consisting of only one breed. They also discovered the same effect through their studies involving simulations.

Cooperation of Breeding Organizations: Because of the high expense of genotyping, it is difficult for a single breeding organisation to develop the large reference population that is necessary to ensure a high level of GS accuracy. As a result, it is feasible that breeding organisations may come to an agreement to work together. The ongoing combined GS effort in North America (VanRaden et al 2009) and Europe (EuroGenomics) (Lund et al 2010) has shown that such a partnership can be beneficial to all parties involved. Each breeding organisation can keep its independence and the security of its genotyping information, even when working together on a project, due to the fact that the original SNP genotype information is not accessible in a G matrix (Harris et al 2009). Interbull (http://www.interbull.org/) is in the process of developing a brand new technology platform that will incorporate data gathered from a variety of sources. This platform ought to make it possible for all parties to benefit from this initiative without having to be concerned about the disclosure of confidential business information. In addition, Daetwyler et al. 2007 demonstrated that GS can simultaneously slow down the rise of inbreeding while boosting the precision of selection in contrast to the conventional BLUP selection. This was done by comparing the two methods side-by-side. The primary reason for this is because when the Mendelian sampling deviation is accounted for in GS, the disparities between siblings are magnified. This is how the GS method works. Inbreeding will be reduced as a result of this since the likelihood of siblings being picked alongside one another will be reduced.

Generation Interval: Using GS, the typical generation time for dairy cattle breeding programmes is reduced from 5–6 years to 1–1.5 years (Pryce and Daetwyler, 2012). Bull testing expenses may be lowered significantly when GEBVs become available at birth, which could accelerate genetic improvement and open up new opportunities (Schaeffer, 2006). Since only male animals are being genotyped and semen of bulls is commonly transported across many nations, the costs associated with genotyping and implementing GS in dairy cattle are lower.

Challenges for Implementation of GS: Although in-depth research on GS has been conducted, and reports of its effective adoption in breeding techniques have been made, this is a relatively new topic, and there are still many obstacles to overcome.

- **Computational Challenges**
  - There are now more SNP markers available as a result of the development of higher density SNP panels such as Illumina's bovine 800K SNP panel, new high-throughput genotyping technologies, and the use of next generation sequencing technology to genetic study. However, as the number of SNP markers increases, the computational requirements become significantly more stringent. For example, if an 800K SNP panel was used for GS instead of a 54K SNP panel, the memory requirement for RRBLUP and the computing time for the Bayesian methods would increase by approximately 200 and 15 times, respectively. This would render the use of larger panels impossible and make it impractical to use larger panels. Multiple trait phenotypes and repeated measurements are routinely recorded, and they are perfectly suited for use in the traditional BLUP model to evaluate the genetic merit of each selection candidate using a repeatability model, a random regression model, or a multi-trait model. This is done by using a combination of a multi-trait model, a random regression model, and a repeatability model. However, only data from a single trait and a single record can be fitted to the current generation of GS models, and this is especially true for Model (1). The GS approach that uses repeated measurements and numerous attributes merits additional examination.

- **Marker-Derived Relationship Matrix**

Even though the dimension of the equations linked to the genetic effects could be lowered from m to n by utilizing the direct GS methods based on Model (2), there are still hurdles that need to be overcome in the development of a marker-derived relationship matrix and in the calculation of its inversion. The numerator relationship matrix is a scattered matrix, and its opposite can be easily built from a pedigree using the information contained in the numerator. On the other hand, the marker-derived connection matrix is a dense matrix, and the principles for generating its inverse are not yet known to anyone. In this regard, the importance of novel rules or parallel arithmetic may not be overstated.

• Genomic Selection in Species other than Cattle

In the dairy cattle sector, genomic selection has seen widespread application in both national and international genetic evaluation. However, the breeding of individuals of other species (especially males) is not nearly as essential in terms of economic significance as the breeding of dairy bulls. There is a possibility that GS will not be cost viable in other livestock species due to the high cost of genotyping as well as the cost of HD panels (about US$200 per sample). The availability of low density SNP panels would result in a reduction in costs and would make it possible for other species to reap the benefits of the promising GS approach (Habier et al 2009, Zhang et al 2011, Weigel et al 2009).

When applied to species other than dairy cow, the implementation of GS faces a number of challenges, one of which is the potential unavailability of highly reliable EBVs. When it comes to dairy cattle, bulls that have EBVs that are known to be very trustworthy can be used to create accurate GEBVs; these accuracies can be determined by calculating the correlations that exist between the GEBVs and the EBVs. It is possible that equivalent data for the males of other species are not currently available; hence, a bigger reference population will be essential in order to achieve the level of accuracy desired for GEBVs. In addition, it is not possible to conduct a direct validation of the performance of GS in these groups.

Cost

In most species, the cost of genomic selection still limits its extension. Many breeding schemes cannot afford the investment to create a reference population, and genotyping costs can still hamper practical implementation: a critical parameter is the ratio of the genotyping over phenotyping costs.

• Genomic Selection Efficiency

It has been observed that genomic selection efficiency is strongly dependent on the relationship between the selection candidates and the reference population.

For a given trait, if the genetic relationships at causal polymorphisms were known, selection efficiency would be maximal. In practice, only a proxy based on genetic markers is used. This proxy is more accurate when relationships are higher, and becomes very poor for almost unrelated individuals. In addition, the marker effects estimated in the reference population reflect the marker–QTL association in this population. When the number of generations separating this population from selection candidates increases, recombination events accumulate between the QTL and their surrounding markers, leading to a loss in efficiency. Therefore, it is anticipated that genomic evaluation would be more robust to lack of close relationships if causal variants or very close markers would be used. For the exact same reason, present genomic prediction methods are not efficient for across-breed selection. A prediction equation built in one breed has nearly no predictive ability in another breed (de Roos et al 2009), except when breeds are closely related. This was initially interpreted as a too low marker density leading to an observed QTL-markers linkage disequilibrium in the reference population, which was not conserved in the candidates. But using a much higher density (with a HD chip with one marker every 4 kb or 0.004 cM) did not improve much the situation.

Conclusion

Genomic selection is an exciting new advancement in animal breeding that aims to increase production by utilising molecular genetic markers to construct unique breeding programmes and to establish new markers-based models for genetic evaluation. This will be accomplished through the use of genomic selection. It paves the way for new research prospects, as innovative algorithms and laboratory procedures are established. The process of genomic selection is applicable to a wide variety of breeds and species. Not only is it desired for the public good, but also for the benefit of the corporate sector, that additional research be conducted on the application of genomic selection (GS) in breeding programmes (breeding companies). It is anticipated that this method will result in an improvement in selection procedures, in particular for species with extended reproductive cycles, late or sex-limited trait recording, or expensive trait recording, and for features that are complicated. The process of incorporating GS into already-established breeding programmes, on the other hand, is not an easy one. In spite of the fact that genomic information has been successfully incorporated into dairy cattle breeding programmes, it has not yet been demonstrated how much weight can be placed on genomic information and how much additional phenotypic information is required from new selection candidates. Genomic selection is already an integral part of the long-term strategy of a great number of breeding companies, including those specialising in the production of pork and beef cattle, to name just two examples. However, additional investigation is required in order to fully estimate how accurate the utilisation of genomic information will be for the prediction of the performance of future breeding stock. The inability to accurately forecast genomic production in crossbreeding and across-breed schemes, the high expense of genotyping, and the limited number of individuals available for testing all contribute to a reluctance to depend solely on genomic information when making selection decisions. The goals of breeding are heavily influenced by the industry, and it is essential to give serious consideration to any additional benefits that may result from employing genomic information. To be effective, this approach is very demanding in terms of both number of individuals genotyped and number of markers on the genome. Its application was unfeasible until the development of large-scale and cheap genotyping methods.

References

Akanno, E. C., F. S. Schenkel, M. Sargolzaei, R. M. Friendship and J. A. B. Robinson. (2013). Opportunities for genome-wide selection for pig breeding in developing countries. J. Anim. Sci. 91: 4617–4627.

Berry, D. F. Kearney and B Harris. (2009). Genomic selection in Ireland. Inter. Bullet. 39: 29.–34.

Boichard, D., Ducrocq, V., Croiseau, P., Fritz, S. (2016). Genomic selection in domestic animals: Principles, Applications and perspectives. C.R.Biologies. 359: 274-277

Daetwyler, H. D., J. M. Hickey and J. M. Henshall, (2010). Accuracy of estimated genomic breeding values for wool and meat traits in a multibreed sheep population. Anim. Prod. Sci. 50: 1004–1010.

de Roos, A. P. W., C. Schrooten and E.Mullaart. (2009). Genomic selection at CRV. Inter. Bullet. 2009, 39: 47–50.

Goddard, M.E. and B.J. Hayes. (2007). Genomic selection. J. Anim. Breed. Genet. 124: 323.–330

Goddard, M.E. and B.J. Hayes. (2009). Mapping genes for complex traits in domestic animals and their use in breeding programmes. Nat. Rev. Genet. 10: 381–391.

Gonzalez, R.O., D. Gianola and G. J. Rosa.(2009). Genome-assisted prediction of a quantitative trait measured in parents and progeny: Application to food conversion rate in chickens. Genet. Sel. Evol.41: 3-10.

Habier, D., J Tetens and F. R. Seefried. (2010).The impact of genetic relationship information on genomic breeding values in German Holstein cattle. Genet. Sel. Evol. 42: 5-10.

Habier, D., R .L. Fernando and J. C. M. Dekkers. (2009). The impact of genetic relationship information on genome-assisted breeding values. Genet. 177: 2389–2397.

Hayes, B. J., P. J. Bowman and A. J. Chamberlain (2009). Genomic selection in dairy cattle: Progress and challenges. J. Dairy. Sci. 92: 433.–443.

Henderson, C. R. (1975). Best linear unbiased estimation and prediction under a selection model. Biometrics 31, 423–447.

Ibanz, N., R. L. Fernando and A .Toosi. (2009). Genomic selection of purebreds for crossbred performance. Genet. Sel. Evol. 41: 12-15.

Kizilkaya, K., R. L. Fernando and D. J. Garrick. (2010). Genomic prediction of simulated multibreed and purebred performance using observed fifty thousand single nucleotide polymorphism genotypes. J. Anim. Sci. 88: 544–551.

Konig, S, H. Simianer and A. Willam.(2009). Economic evaluation of genomic breeding programs. J. Dairy Sci. 92: 382.–391

Lillehammer, M., T. H. E. Meuwissen and A. K. Sonesson. (2011). A low-marker density implementation of genomic selection in aquaculture using within-family genomic breeding values. Genet. Sel. Evol. 45: 112-116.

Luan, T, J. A. Woolliams and S. Lien. (2009). The accuracy of genomic selection in Norwegian red cattle assessed by cross-validation. Genet.183: 1119–1126.

Lund, M. S., and G.Su. (2009). Genomic selection in the Nordic countries. Interbull bullet. 39: 39–42

Meuwissen, T. H. E., B. J. Hayes and M. E. Goddard (2001). Prediction of total genetic value using genome-wide dense marker maps. Genet.157: 1819–1829.

Schaeffer L. R. (2006). Strategy for applying genome-wide selection in dairy cattle. J. Anim. Breed. Genet. 123: 218–223.

Schefers, J. M., andK. A. Weigel (2012). Genomic selection in dairy cattle: integration of DNA testing into breeding programs. Anim. Front. 2: 4–9.

Toosi A., R. L. Fernando and J. C. Dekkers (2010). Genomic selection in admixed and crossbred populations. J. Anim. Sci. 88: 32–46.

Tribout, T., C. Larzul and F. Phocas (2012). Efficiency of genomic selection in a purebred pig male line. J. Anim. Sci. 90: 4164–4176.

Weigel, K. A., G . de los Campos and O. Gonzalez-Recio (2009). Predictive ability of direct genomic values for lifetime net merit of Holstein sires using selected subsets of single nucleotide polymorphism markers. J. Dairy Sci. 92: 5248–5257.

Williams, J. L. (2005). The use of marker-assisted selection in animal breeding and biotechnology. Rev. Sci. Tech. 24 : 379–391.

Wolc A, C. Stricker and J. Arango (2011). Breeding value prediction for production traits in layer chickens using pedigree or genomic relationships in a reduced animal model. Genet. Sel. Evol. 43: 5-8.

Zhang, Z., Q. Zhang and X. D. Ding. (2011). Advances in genomic selection in domestic animals. Chin.Sci. Bull. 56: 2655-2663

Zhang Z., X. D. Ding and J. F. Liu (2011). Accuracy of genomic prediction using low density marker panels. J. Dairy Sci. 96: 2011-2018.

21

Better Animal Husbandry and Animal Breeding with Application of Artificial Intelligence in Biostatistics

Rakesh Goel[1], Avneesh Kumar[1], S.P. Singh[1], Deepak Sharma[1] and Neha Gupta[2]

[1]Department of Animal Genetics & Breeding, C.V.Sc. & A.H., DUVASU Mathura, Uttar Pradesh
[2]Department of Animal Husbandry, Government of Uttar Pradesh, Uttar Pradesh

---

Biostatistics is a branch of statistics that focuses on the analysis and interpretation of data from biological and health-related research studies. It involves the application of statistical methods to design, collect, analyze, and interpret data from studies involving living organisms like: animals, birds including human beings. Biostatistics plays a crucial role in public health, medicine, epidemiology, genetics, and other related fields.

With the increasing availability of data and advancements in biostatistics, breeders can expect further improvements in breeding practices that will results in better outcomes for both livestock farmers and the animals they breed. Animal breeding is a crucial aspect of agriculture and food production, as it helps to improve the quality and quantity of animal products as well ensuring food security to the masses.

Importance of Biostatistics

Biostatistics is important for several reasons:

Designing studies: Biostatistics is essential for designing studies that are well-controlled minimize bias, and it is essential for generating reliable and accurate results. It is a critical component of biological and health-related research. The importance of biostatistics can be elucidated with following points-

1. Analyzing data: Biostatistics provides tools and techniques for analyzing complex biological and health-related data, which can reveal patterns, trends, and relationships that are difficult to detect otherwise.
2. Drawing conclusions: Biostatistics allows researchers to draw meaningful conclusions from data, and to make inferences about populations based on samples.
3. Public health: Biostatistics plays a crucial role in public health by enabling the identification of risk factors for diseases, the evaluation of the effectiveness of interventions, and the monitoring of health outcomes.
4. Precision medicine: Biostatistics is essential for developing personalized treatment plans based on individual characteristics and health data.
5. Resource optimization: Biostatistics allows researchers to optimize the use of resources, such as time, money, and personnel, by designing studies that generate the most information with the fewest resources.
6. Decision-making: Biostatistics provides a scientific basis for decision-making in biology and medicine, allowing researchers, clinicians, and policymakers to make informed decisions based on empirical evidence.
7. Quality control: Biostatistics is crucial for ensuring the quality and validity of research studies, by providing methods for identifying and controlling sources of bias and error.
8. Data-driven policy: Biostatistics can inform health policy decisions by providing data-driven insights into the prevalence, incidence, and distribution of diseases and health conditions.
9. Innovation: Biostatistics can drive innovation in biology and medicine by providing the tools and techniques necessary for exploring new areas of research and developing new treatments and interventions.

Artificial Intelligence (AI)

It refers to the ability of computers or machines to perform tasks that would normally require human intelligence, such as recognizing images, understanding natural language, making predictions, and learning from data. AI systems use algorithms and mathematical models to analyze data, find patterns, and make decisions or predictions based on that data. In essence, AI involves creating machines that can perform tasks that would normally require human intelligence, such as problem-solving, decision-making, and learning.

Importance of Artificial Intelligence (AI) in Current Scenario

AI has the potential to transform many aspects of our lives and improve productivity, innovation, personalization, safety, and sustainability. Hence the importance of artificial intelligence (AI) in the current scenario can hardly be overstated. Following are some reasons why AI is so important today:

1. Efficiency: AI has the potential to greatly increase efficiency in many industries, automating tasks that were previously performed manually and freeing up time and resources for more important tasks.
2. Innovation: AI is driving innovation in many fields, from healthcare and education to finance and transportation. AI is helping to solve complex problems, develop new products and services, and discover new insights.
3. Personalization: AI is enabling more personalized experiences for consumers, from targeted advertising and recommendations to customized healthcare and education.
4. Safety: AI is improving safety in many industries, such as transportation and manufacturing, by identifying potential risks and providing early warning systems.
5. Sustainability: AI is helping to promote sustainability by improving resource management, reducing waste, and enabling more efficient use of energy.

Artificial Intelligence (AI) and Biostatistics at a Crossroads

The use of artificial intelligence (AI) in biostatistics can be traced back to the early days of AI research in the 1950s and 1960s. However, the application of AI to biostatistics was limited by the availability of data and computational resources, as well as by the limitations of early AI methods.

In recent years, with the explosion of data and advances in computing power, AI has become increasingly important in biostatistics. Machine learning algorithms, in particular, have been widely used to analyze and interpret biological and health-related data, and have led to many new discoveries and insights in the field.

Overall, while the roots of AI in biostatistics can be traced back several decades, the recent advances in data and computing have greatly expanded the potential for using AI in biostatistics research by enabling the processing and analysis of large and complex datasets. With more powerful computing resources, biostatisticians and artificial intelligence algorithms can handle increasingly large amounts of data, extract insights from it, and develop more accurate models and predictions. Additionally, the availability of massive storage capacities means that large datasets can be stored and easily accessed for future analysis, making it easier for researchers to work with vast amounts of data. This has allowed for more sophisticated statistical models and machine learning algorithms to be developed and applied to a wide range of problems in biology, animal breeding and in field of medicine.

Role of Artificial Intelligence and Biostatistics in Animal Husbandry: The application of artificial intelligence is having an effect on every facet of life, and it also has the potential to completely transform the field of animal husbandry. The following are a few instances of how AI can be used to fuel the field of animal husbandry:-

1. In Monitoring and Analyzing Animal Behavior

One key application of artificial Intelligence in animal husbandry is the use of sensors and cameras to monitor animal behavior and detect changes in behavior that may indicate illness or other health problems. Artificial Intelligence algorithms can analyze data from cameras to detect changes in an animal's movements, eating patterns, and other behaviors that may indicate pain, distress, or the onset of disease. This information can then be used to quickly diagnose and treat any health issues and reduced losses due to illness or death. To understand how above aim can be achieved we need to think about requirements of hardware, software, and model of AI needed for monitoring and analyzing animal behavior can vary depending on the specific application and goals of the project. However, in general, some common elements that may be required include:

i. Hardware: Cameras, sensors, and other devices for capturing video or other data on animal behavior. The specific hardware needed may depend on the species and behavior being studied.

ii. Software: Software for processing and analyzing the data collected by the hardware. This can include software for video analysis, statistical analysis, and machine learning.

iii. Model of AI: A model of AI, such as a machine learning algorithm, can be used to detect and classify animal behaviors automatically. This can save time and effort by automating the analysis process.

iv. Examples of specific hardware, software, and AI models that may be used for monitoring and analyzing animal behavior include:

v. Hardware: High-speed cameras, accelerometers, GPS trackers, microphones, and other sensors

vi. Software: OpenCV, MATLAB, EthoVision, Noldus Observer, and other software for video analysis and statistical analysis

vii. Model of AI: Convolutional neural networks (CNNs), recurrent neural networks (RNNs), support vector machines (SVMs), and other machine learning models.

2. In Monitoring Health of Farm Animals

Artificial Intelligence (AI) is playing an increasingly important role in animal husbandry, enabling farmers and ranchers to improve the health and well-being of their livestock. The specific algorithm used for diagnosing diseases in farm animals with AI can vary depending on the type of animal, the disease being diagnosed, and the available data. Common algorithms used in this context include decision trees, random forests, and neural networks. The process of diagnosing diseases in farm animals with the help of AI typically involves the following steps:

i. Data collection: Collecting data about the animals, such as their breed, age, weight, and any observed symptoms, as well as data about their environment, such as temperature, humidity, and lighting.

ii. Data pre-processing: The collected data is pre-processed to remove any irrelevant or redundant information, and to normalize the data so that it can be used effectively by the AI algorithm.

iii. Feature extraction: Features are extracted from the pre-processed data, which are used to represent different aspects of the animal's health, such as its temperature, heart rate, and respiratory rate.

iv. Algorithm training: The AI algorithm is trained using a large dataset of labeled data, which contains information about which animals have which diseases. The algorithm learns to recognize patterns in the data that are associated with different diseases.

v. Diagnosis: Once the AI algorithm is trained, it can be used to diagnose diseases in new animals by inputting their data and analyzing it using the learned patterns. The algorithm can then output a diagnosis, which can be verified by a veterinarian.

3. In improvising Treatment of Diseases in Farm Animals

AI can be used in the treatment process of diseases in farm animals by developing algorithms that can analyze various data sources, such as animal health records, environmental conditions, and feed quality, to detect early signs of illness and predict disease outbreaks. The algorithms can then provide recommendations for treatments and preventive measures based on the specific needs of each animal. Additionally, AI can be used to optimize the administration of medications, including dosage and frequency, by analyzing factors such as the animal's weight, age, and health status. By continuously analyzing data, the algorithm can learn and improve its accuracy over time, leading to more effective and efficient treatments for farm animals.

4. In Optimizing the Feed and Nutrient Distribution

AI can be used to optimize feed and nutrient distribution in a livestock farm by developing algorithms that can analyze data on the nutritional requirements of each animal, the composition of available feed, and environmental conditions such as temperature and humidity. The algorithms can then recommend the optimal feed formulation for each animal based on their age, weight, and health status, as well as the amount and timing of feed distribution.

The algorithm can also consider factors such as the cost of feed ingredients and the availability of different feed sources to develop the most cost-effective feeding plan. The algorithm can also learn and adapt to changing conditions and adjust the feeding plan accordingly leading to improved animal health, increased productivity, and reduced feed waste.

Furthermore, AI can also be used to monitor the feeding behavior of animals so that the feeding plan can be adjusted accordingly. If an animal is not eating as much as usual, the artificial intelligence can adjust the feeding plan to ensure that the animal receives the necessary nutrients. This can prevent under nutrition and improve the overall health of the animal. The AI model can be trained continuously according to learning and adapting to changing conditions.

5. In Automating Many of the Manual Processes in a Livestock Farm

AI can automate various processes in a livestock farm; there are several processes in a livestock farm that can be automated with the help of AI, some of which are mentioned as following:

i. Feeding: AI can be used to develop personalized feeding plans for each animal, monitor feeding behavior, and adjust the feeding plan accordingly.

ii. Health monitoring: AI can analyze various data sources, such as animal health records, environmental conditions, and feed quality, to detect early signs of illness and predict disease outbreaks.

iii. Breeding: AI can analyze genetic data to optimize breeding plans and increase the quality and productivity of the herd.

iv. Reproduction: AI can monitor the reproductive behavior of animals and provide recommendations for breeding based on factors such as age, weight, and health status.

v. Growth monitoring: AI can monitor the growth of animals and provide recommendations for feed and nutrient distribution to optimize growth and productivity.

vi. Environmental control: AI can monitor environmental conditions such as temperature, humidity, and air quality, and adjust environmental controls such as lighting and ventilation to optimize animal comfort and productivity.

vii. Record-keeping: AI can automate the collection and analysis of data such as animal health records, feed usage, and growth rates, to help farmers make informed decisions about herd management and resource allocation.

6. In Improving Animal Growth and Development

Artificial Intelligence algorithms can analyze data on animal growth and development but several steps must be taken to tap potential of artificial intelligence for improving animal growth and development. Care must be taken in data collection and analysis on the genetic background, environment, and health status of the animals. Critical elements influencing animal growth and development, such as diet, disease prevention, and environmental conditions should be identified. To create a prediction models that takes these critical parameters into consideration and forecasts the most effective interventions for encouraging animal growth and development. Implementing the algorithm and tracking its success over time in a farm management system. Efficacy, of the algorithm is constantly refined based on new data and insights.

7. In Boosting Profits and Cutting Losses in Livestock Enterprise

The use of artificial intelligence can provide real-time monitoring and analysis of various aspects of animal care, such as feed intake, health status, and environmental conditions, can increase the profitability of animal husbandry while simultaneously reducing waste and losses. This is because AI can monitor and analyze various aspects of animal care in real time. Utilizing AI technology allows for the collection and examination of data, which can then be used to optimize the amount of feed and water consumed, forecast and prevent disease outbreaks, and increase the overall efficiency of the process of producing animals for consumption. In addition, AI can assist in the formulation of individualized treatment programmes and in the decision-making process pertaining to breeding and the selection of genetic material.

8. In Boosting Efficiency of Animal Husbandry Operations

In addition to providing for the welfare of animals, artificial intelligence has the potential to increase productivity in animal husbandry by automating a variety of jobs. These tasks include the gathering of data, conducting analysis, and making decisions. For instance, artificial intelligence can be used to improve the scheduling of farm activities like feeding, cleaning, and maintenance based on parameters such as the weather, the behaviour of animals, and the availability of resources. Monitoring and managing the supply chain, including inventory management and distribution logistics, can also be made easier with its assistance. In addition, sensors and cameras that are powered by AI may be used to keep an eye on the physical infrastructure of the farm, such as the barns and the machinery, to ensure that it is in excellent shape and operating as efficiently as possible. In animal husbandry as a whole, artificial intelligence has the potential to assist streamline processes, cut expenses, and boost output.

9. Promoting Innovation in Livestock Farm Practices

Artificial intelligence has the potential to stimulate innovation in cattle farming techniques by providing insights that are data-driven and optimizing a variety of farming processes. The use of sensors and cameras can make it possible to monitor animal health and behaviour, and they can also provide an early warning in the event that there is a potential problem with the animal's health. It is possible to do research on environmental factors such as temperature, humidity, and illumination in order to devise environments that are best for the development and production of animals. The formulation and delivery of feed can be optimized, resulting in an increase in the nutritive value of the feed as well as a decrease in the amount of waste produced. The use of AI allows for the prediction and management of disease outbreaks, which helps to reduce the transmission of infectious illnesses. It is possible to automate processes like milking, cleaning, and feeding, which leads to higher efficiency and lower expenses associated with labour.

10. Role in Animal Breeding

Biostatistics plays a crucial role in animal breeding, as it provides a systematic and scientific approach to analyzing data related to animal populations. This includes data on genetics, physical and behavioral traits, as well as health and performance indicators. By using statistical methods and models, biostatistics provides valuable insights that can inform and improve breeding decisions.

AI has the potential to improve animal breeding by combining the disciplines of genetics, biostatistics, and computer science. This will make it possible to analyze vast datasets and determine which animals possess desirable characteristics. Geneticists are now able to detect genetic markers associated with qualities such as disease resistance, milk yield, and meat quality through the use of artificial intelligence. Biostatisticians have the ability to construct models that can examine genetic data and make predictions about how future generations of animals will perform. The huge amounts of data that are produced by genetic and bio-statistical investigations can be processed and analyzed using algorithms that can be developed by computer scientists.

i. **Marker Assisted Selection-** Artificial intelligence (AI) facilitates the study of enormous amounts of genetic data, which aids in marker-assisted selection (MAS) and the identification of genes associated with disease resistance. Animals that have desirable features, such resistance to disease, can be selected for breeding by analyzing their DNA and looking for certain genetic markers. Large genomic datasets from animals with known characteristics can be analyzed by AI to help locate these markers. In the end, this can speed up the procedure of discovering and selecting animals with suitable qualities for breeding.

ii. **Production of Disease Resistant Animals-** AI can also be used to identify genes associated with disease resistance by analyzing genetic data from animals that are resistant to a particular disease. By comparing the genetic data of resistant and susceptible animals, AI can identify genes that are associated with disease resistance which further can be used to develop breeding programs that select for animals with resistance to specific diseases.

iii. **To finds Patterns and Connections in Large Datasets from Multiple Sources and Predicts Future Trends and Patterns-** Artificial intelligence (AI) has the potential to be very helpful in quickly detecting patterns and correlations in enormous datasets from many sources, such as DNA sequencing, and extracting insights from these databases. Using machine learning algorithms, AI can sift through mountains of genetic data in search of patterns and correlations that humans might miss. The time and effort spent on study can be reduced, and potentially new genetic correlations can be found, with this. Artificial intelligence can make predictions about future trends and patterns by analyzing past data. By analysing massive amounts of genetic data, AI can discover patterns and relationships that are indicative of future occurrences. This can help breeders find animals with desirable traits, forecast how future generations will do, and come up with novel approaches to breeding that improve animal populations' productivity and resilience. To further our understanding of animal biology and performance, AI can be used to incorporate data from multiple sources, such as genetic and environmental data. By analyzing large datasets from a variety of sources, AI can discover complex relationships between genetic and environmental factors, shedding light on the factors that affect animal health and performance.

iv. **Prevention of Inbreeding** -Inbreeding can be minimized in livestock breeding programs by developing an AI algorithm that would need analyze the genetic data of animals by following procedure.

By application of biostatistics and bioinformatics an algorithm need to be developed to examine the genetic data of animals and determine which ones are the least connected to one another so that inbreeding may be avoided in livestock breeding. The algorithm would need to take into consideration a number of different elements, including the genetic markers that are related with specific behaviours, the frequency of those genetic markers in the population, and the pedigree information of the animals. Utilizing a strategy known as "optimal contribution selection" is one method that can be utilized to prevent inbreeding. This involves assessing the genetic data of potential parents and choosing those who will produce offspring with the greatest possible genetic diversity while reducing the likelihood of inbreeding as much as possible. In order to put this strategy into action, a computer programme might be designed to examine the genetic information of prospective parents and apply optimization principles in order to determine which pairs of animals would make the most compatible partners. Gene dropping simulation is another strategy that can be utilized to reduce the likelihood of inbreeding occurring. This entails simulating the genetic transmission of alleles from parents to kids and predicting the chance of inbreeding based on the genetic relatedness of the possible parents. Inbreeding can occur when two individuals who are genetically related have children. By simulating the genetic transfer of alleles and making use of statistical methods to forecast the risk of inbreeding, an AI system might be constructed to put this method into practice to minimize the animal breeding.

Conclusion

The animal husbandry sector and in particular animal breeding both are going to have various new implications as AI is being applied, but AI will make it possible to conduct data analysis and predictive modeling with a higher degree of precision than was previously possible without the need for active human involvement.

It is evident from above information presented in this chapter that like every aspect of day to day life Artificial intelligence is going to impact animal husbandry field also with good effect their might also be some repercussions by the application of artificial intelligence in the field animal breeding and animal husbandry, and their might be potential unintended consequences. These consequences may include the requirement for adequate data management as well as ethical considerations related to the use of animal data. Concerns may also exist around the potential loss of jobs and skills connected to more conventional manual techniques of data processing. In order to make responsible and productive use of artificial intelligence in this area, it is essential to thoroughly evaluate and handle the various implications that could arise from its application.

It can be concluded that the fields of animal breeding and animal husbandry both stand to benefit from the application of biostatistics and artificial intelligence (AI).

22

The Impact of Artificial Intelligence on Animal Breeding Research

Avneesh Kumar[1], Rakesh Goel[1], S.P. Singh[1], Deepak Sharma[1] and Neha Gupta[2]

[1]Department of Animal genetics & Breeding, College of Veterinary Sciences and Animal Husbandry, DUVASU, Mathura
[2]Department of Animal Husbandry, Government of Uttar Pradesh

The term "artificial intelligence" (AI) refers to a wide variety of theories and technologies that are used to the resolution of issues characterized by a high level of logical or algorithmic complexity. It cuts across a wide range of fields, such as mechanical modeling, software engineering, data science, and statistics. Many artificial intelligence (AI) technologies, which first appeared in the 1950s, have lately been refined or extended thanks to advances in computer performance. Recent advancements have been made possible by the creation of interfaces between artificial intelligence and other fields of study, such as biomedicine, and by the collection of enormous amounts of data from a variety of fields, notably those related with the healthcare industry. First, Artificial intelligence helps with situational awareness, which is relevant in animal health and animal husbandry because it allows for the detection of patterns like repeated sequences of observations, forms like a protein, and signals like increased mortality especially in comparison to a baseline at various scales. A third option is computer-based decision making, or more practically, human decision support via knowledge - based systems, diagnostic support, and resource allocation, for instance.

The field of artificial intelligence (AI) develops many different ideas and strategies to address these problems. One such well-known AI technique is machine learning (ML), which has been evolving since the 1980s (Murphy et al., 2012). Since the turn of the century, deep learning has progressed alongside the proliferation of big data and the ever-increasing power of computers, allowing for the investigation of data sets too large to be processed by traditional statistical approaches. Methods and algorithms for addressing complicated problems, automating processes or reasoning, integrating data from several sources, and providing decision assistance all fall under this category. Although these techniques are on the rise in human medicine, they are infrequently applied to the research of animal breeding scenarios.

Some of the scientific challenges that are faced in animal breeding can also be addressed from a fresh angle by employing some of these AI methods to evaluate the ever-increasing data that is collected on animals' production, reproduction, and their environment. This can help solve some of the problems that have been plaguing the field of animal breeding for quite some time. Research on animal breeding can benefit from developments in machine learning and deep learning methods, such as those used in transcriptomics, genomics, and to examine proteomics, among other areas. These strategies lead to enhanced selection, as well as more reliable forecasts and fewer errors; accelerated decision making; greater accuracy in breeding value analysis; and improved intervention targeting in animal breeding. The study of animal breeding can also benefit from scientific discoveries made in other areas of artificial intelligence. The modelling of reasoning and the representation of knowledge make it possible to create more accurate representations of complicated sociobiological systems, such as those that are faced in animal breeding. Some examples of such processes are those pertaining to decision-making and the management of uncertainty, as well as the life course of breeding females. This helps to make things more readable for people who are not trained in computer science. In addition, developments in problem solving under constrained resource allocation, autonomous agents, multi-agent systems, and multi-level systems, as well as advancements in automatic computer code generation, can be utilised to improve breeding models' levels of efficiency and reliability. It's interesting to note that this might help in predicting how decisions about management and control would play out at various temporal and spatial scales (animal, herd, country etc.)

Research conducted at the interface of animal breeding and artificial intelligence (AI) leads to the identification of new difficulties for AI. These challenges are on topics that are common to population studies, but they examine distinct contexts and perspectives. When dealing with animal breeding, the first and most important step is to ensure that the specific agro- and socio-economic conditions of production systems are taken into consideration. Human actions and choices are essential to the functioning of animal production systems. In family farming, they can be a source of money (for example, through cattle), a source of labour forces, and a supply of food all at the same time. Farmers also have high standards in regards to ethical behaviour and the care of their animals. There is a possibility that conventional methods of animal breeding would no longer be employed with advent of genomic technologies. AI can help in the process of identifying and evaluating potential alternatives, which will require the mobilisation of both AI technologies and new animal breeding data streams. These data will be distinct from the data used in population research. The integration of data from deep sequencing into animal breeding is a problem in and of itself. This includes the use of developing technologies for the investigation of the metabolome and epigenome. Second, the management of breeding programmes and animal populations are both influenced by the recurring decisions that regulate animal populations (e.g., trade, control measures). Because of this, economic criteria, which are the implications of animal breeding techniques on the incomes of livestock farmers, are crucial indicators for evaluating animal breeding strategies, which can sometimes be misunderstood or may be in conflict with the expectations of society. Because of these specifics, the Animal breeding/AI interface has become a topic of interest as a way to drive new methodological work and to solve some of the ancient and contemporary problems that are being encountered in Animal breeding research today. With the development of omics technologies such as transcriptomics, epigenomics, and proteomics, fostering multi-disciplinarity, stakeholders' participation, data sharing, and addressing the intricacy of breeding concerns (for example, loss of genetic diversity, short- and long-term climatic influences on animal husbandry practices), artificial intelligence could assist in this new development by allowing it to technically tackle some of the difficult problems addressed.

This chapter discusses a variety of topics pertaining to data collection, organization, and access; how Methodologies make a contribution to reconsidering our understanding of traditional breeding; how AI methods facilitate decision-making and induce innovation; and possible barriers to the advancement of artificial intelligence in modern breeding, as well as the levers that can be used to overcome those barriers. We have also looked into the recommendations that can help us meet the new issues that have been brought on by the interaction between animal breeding and AI interface.

Acquire, Sort, and Make Available Relevant Information

The availability of high-quality data at various organizational levels of biological systems and, as a consequence, across a wide range of temporal and spatial scales continues to be an important focus of research in the field of animal breeding. There are many different kinds of data of interest. They can be gathered through the use of molecular analysis (for example, genomic, metagenomics, or metabolic data), by the collection of observational data on individuals (for example, temperature, behavior, dairy production and composition, weight, and feed intake), or through the production system for example herd structure, breeding, management of cleanliness etc). It is also possible to acquire them on a bigger basis, going past herds or groupings of livestock for example commercial movements.

Although it is difficult to acquire data of this massive and varied nature (for example, genomic data), a significant and varied amount has already been gathered in the following ways: first via reporting requirements in accordance with laws (for example commercial movements of livestock, surveillance framework); secondly by intuitive devices (e.g., sensors, surveillance cameras systems); and lastly on to an ad hoc basis as part of the study programmes. This results in a very wide variety of data attributes, which in turn results in a very wide variety of their management, access, and prospective uses. These data can be gathered in a targeted manner for certain herds or animals (for instance, through cohort monitoring programmes) or by private businesses (for example milk collection). Because of this, access to academic research and public research may be restricted. Because of globalization and large-scale livestock trading, there may be a need to use data gathered at a worldwide scale in animal breeding. This is especially true for quantitative genetics, which includes areas such as animal genetics and breed management. This may cause problems with standardization.

It is important to give some thought to potential future methods for monitoring, collecting, and maintaining these data, as well as to practices that aim to improve collaboration amongst various stakeholders. It has become a strategic priority in both theoretical and more applied research, aided with a technological and algorithmic challenge (Neethirajan et al 2017). Data management was always an important component in applied research to facilitate the use and valorization of the data, but it has recently evolved into a more prominent role as a strategic issue. In point of fact, developing algorithms that can effectively manage enormous data flows and stocks by optimizing calculations is a difficult task, especially when done in real time. It appears to be important, as well, to make disparate data sources compatible, which will need the development of specific analytical approaches. A significant portion of the data is also confidential, and its ownership is frequently complex (for example, there may be multiple owners, the data may not be centralized, and the data may be closed), and it is occasionally ambiguous (for instance, there may be a lack of knowledge regarding the true owner of the data, which may be the farmers, the data collector, or the farmers' union). All of this has a tendency to significantly aggravate access to this information, raises serious questions regarding intellectual property, as well as raises concerns when it comes to legislation with regards to information protection, for example, the adaptation of legislation to Animal breeding while respecting the secrecy of the private information that is mobilized. What is the appropriate business model for the collecting of data or access to the various databases that already exist? What about the accessibility of the AH data (for example, the duality between the concept of public benefit and the proprietary nature of certain data), which allows it to experiment with AI algorithms in real-world settings and evaluate their effectiveness? If you could provide responses to these questions, it would make the collecting and exchange of ad hoc data much easier. Building accurate approximations of complicated socio-biological systems is made easier with the assistance of artificial intelligence (AI), especially when a participatory framework, expert systems, and multi-agent systems are combined. As a result, it is demonstrated to be an efficient instrument for fostering the participation of various stakeholders in the process of collaborative and optimized decision-making, as well as for evaluating the impact of changes in usage and practices (Binot et al 2015).

It is of the utmost importance to foster the testing of artificial intelligence (AI) technologies on a regional scale in order to foster their growth, assess their effectiveness and measure the predictive quality of these technologies. In the field of animal breeding, facilitating easy accessibility to information facilities would make it possible to test creative solutions on a wider scale, which in turn would speed up the process of developing and evaluating those solutions. There is a substantial amount of expertise available that might be put to beneficial use. Some examples of this are breeding data platforms like INAPH, big cohorts, and experimental farms. AI could also be used to help rethink sampling techniques for data gathering in animal breeding. This would allow for improved and more dynamic targeting of the data that needs to be collected, while simultaneously minimizing the collection of redundant, collinear, and unnecessary data.

Artificial Intelligence's Contribution to our Comprehension of Breeding Programmes for Animals

New tools utilizing AI methods have made it possible to collect and store massive amounts of data on measurements and observations. This has increased the need for reliable methods of data analysis. The computer science community responded to these needs by developing AI techniques that take use of the exponential growth in computing power. Computational complexity in the domains of parameters and variables attribute selection, and modeling evaluation and combination are only a few examples of how statistical methods have advanced over the past few decades. The explosion of computing capacity has allowed for the refinement of Bayesian inference using simulated or approximatively methods (Robert CP, 2014). While shifting the paradigm of probabilistic reasoning, Bayesian methods have made it easier to combine data from several sources, account for prior knowledge, and draw conclusions from more complicated and realistic models.

Interventions that Hit Their Mark: Choose Amongst the Options

Animal breeders face the issue of determining which combinations of mating pairings are most relevant to local (such as farm characteristics and production goals) and territorial (such as available resources, farm location, and management priorities) specificities. They are responsible for providing quality breeding guidance in addition to anticipating the effects of changes in health, environmental, and regulatory standards. The question of how to encourage innovation in animal breeding also arises; for example, how to predict the required characteristics of a selected candidate in breeding or how to evaluate the challenging advantage of new strategies (such as genomic selection of resistant animals or new strains) in comparison to more traditional methods. Private managers (for example, farmers and farmers' advisors) and collective managers (for example, farmer groups and public authorities) need decision support tools to more accurately aim general populace incentives, distinguish investment opportunities to be favoured by farmers, and target the metrics as effectively as possible. Who to aim (that what farms, which animals)? These are just some of the questions that need to be answered. Within the context of the agro-ecological transition, these concerns become crucial to the process of deliberating over the utilisation of inputs.

The utilization of mechanistic modelling is an approach that may be utilized to evaluate, analyze, and prioritize ex ante a large variety of available possibilities. However, the vast majority of the models that are currently available do not explicitly incorporate human decision-making, despite the fact that farmers frequently make control decisions (for example, unregulated mating), which can have potentially huge breeding and decision-making repercussions (for example dissemination of information). Recent research has been conducted with the intention of integrating humans and the decisions they make by employing optimum control and adaptive techniques derived from artificial intelligence or farm economics methodologies. To design control strategies that are both transparent and context-sensitive is a difficult task. Such research is just getting started in the Animal breeding field and needs to be expanded as part of the advancement of agro-ecology. This is necessary in order to meet the society's current demand for quality of product and regard for ecosystems and their bio - diversity on the one hand, and animal welfare and ethics on the other.

Obstacles to Progress in Artificial Intelligence and Animal Breeding Interface

Research that is carried out at the interface between AI and animal breeding necessitates robust interactions between biological disciplines (such as immunology, clinical sciences, genetics, ecology, evolution, epidemiology, animal and veterinary sciences) and much more conceptual disciplines (such as modeling, statistics, computer science), sometimes in conjunction with sociology and economics. This work needs to be founded on a wide variety of methodological abilities (for example data mining, information systems, problem solving, meta-modeling, optimization, decision making models). At both the national and the international level, important challenges include the requirement for research, training, and support. In addition, there is a need for a connection that is both facilitated and trusted between academic institutions, institutes of technology, and corporate partners. These entities are frequently the holders or collectors of data that is relevant to animal breeding exploratory questions that can be answered using AI strategies. Because this topic appears to be overly competitive and because some existing divisions continue to be resistant to the sharing of information and data, it is imperative that improvements in intersectoral communication and coordination be developed at a level that is higher than that of individual institutions.

In order to face the difficulties offered by agriculture in the 21st century, it is necessary to suggest a cultural assimilation of scientists to AI, its methodologies and possible breakthroughs, as well as its limitations. Indeed, there are challenges involved in carrying out study along this line of scientific inquiry. Given the small number of academic staff working on these problems, their extremely high current mobilization, and their limited inventory to work collaboratively on new subjects, it is difficult to establish the new collaborations that are needed among team members undertaking systematic work and workgroups inside the fields of application. This is made more difficult by the fact that it is difficult to understand and master these methods. It is necessary to keep track of the various AI approaches that are either already available or in the process of being developed, as well as new software and package releases, and train personnel on how to use them. Interconnectivity can also be formed via transversal teams, which is a favoured path, and this allows for the development of essential collaborations as well as the establishment of a strategic orientation. Find solutions to stimulate technique precipitation in the society and the growth of scientific and technological skills. This is another problem that has to be solved.

Finally, artificial intelligence (AI) methods such as categorization, machine learning, and data mining, as well as the advancements in AH that these methods can lead to, are barely mentioned in veterinarian curriculum, despite the fact that these youngsters represent the future of animal breeders. In a similar vein, there is a lack of sentry groups of veterinarians, despite the fact that these networks are in the process of emerging, despite the fact that questions regarding animal breeding can arise on a broad and communal scale. The research establishment would also gain from further strengthening its knowledge and skills in the valuation, transfer, and safeguarding of intellectual property on such AI technologies and associated outcomes. This would be a win-win for everyone involved.

Indicators of a Successful AI-animal Breeding Interface

i. **Confidentiality of Information Exchange:** Without robust assistance for the administration of data storage, maintenance, assessment, computation, and restitution, there is no way to innovate at the intersection of AI and Animal breeding. This precludes any possibility of advancement in any field. The most significant threat is that there will be an increase in the need for breakthroughs in AI, but there will not be enough human resources to meet that demand. Additionally, a competence in legislation, jurisdiction, and morality is required in order to acquire, hold, utilize, and safeguard data in Animal breeding.

ii. **Bring in the Right Expertise:** The training of the staff is one way to acquire new abilities. Another one is collaborative approach, however, because the trained teams are already at their maximum capacity, beginning cooperation at the Animal breeding/AI interface becomes a highly hard process. Increased skill capitalization in innovative methods, together with the facilitation of future national or worldwide cooperation, could be achieved through the provision of financial incentives for scientific internships in specialized laboratories.

iii. **Promote the Creation of Animal breeding/AI Initiatives:** When working on projects that involve the intersection of animal breeding and artificial intelligence, as with any other type of interdisciplinary work, it is necessary to bring together teams of experts from both fields and give everyone the opportunity to advance in their own fields. However, in order to identify the problems that are shared by the most relevant disciplines, it is necessary to have a good cultural assimilation of the fields of study among them, as well as an otherness that is directed at greater understanding each other (Duboz et al 2018). This is not yet the case at the interface between animal breeding and AI.

iv. Fostering Creativity and Collaboration Between Academia and Industry

If industry-academia partnerships were encouraged, it might be conceivable to permanently move AI R&D to Animal breeding, creating a strong effect on public financing. It would be helpful to map the rapidly evolving ecosystem of startups in the Animal breeding/AI field, regardless of whether they're established multinationals or fledgling start-ups, in order to better comprehend the potential interconnections among them. Academic outputs created at this interface could benefit from greater exposure and attention if they were mapped. Lastly, this more innovate would be bolstered by recognizing the generation of recorded algorithms as academic outputs alongside articles. Further, it would be wise to launch an information and education/acculturation policy concerning AI and its growth in AH.

Summary

Today, AI methods like machine learning, knowledge-based platforms, and computational technologies combine with massive volumes of complex data to accelerate these fields. Big data and AI must not be grouped together. Information collecting doesn't necessarily improve understanding. The more data that represents functional notions and assumptions, the much more useful AI applications can be. To effectively manage multi-sectoral data and their many applications, animal breeding must focus on the moral, categorical, and regulatory aspects of information protection, storage, administration, sharing, and compatibility locally and globally. Even with great data collection, animal breeders' AI abilities are still far behind the needs. Since AI teams are in demand, collaboration is limited. Animal breeders need specific and generalized training to comprehend AI's possibilities and limits. In conclusion, AI allows the many upstream participants in animal breeding and welfare to share their data and perspectives.

References

Binot A, Duboz R, Promburom P, Phimpraphai W, Cappelle J, Lajaunie C, Goutard FL, Pinyopummintr T, Figuié M, Roger FL (2015) A framework to promote collective action within the One Health community of practice: using participatory modelling to enable interdisciplinary, cross-sectoral and multi-level integration. One Health 1:44–48.

Duboz R, Echaubard P, Promburom P, Kilvington M, Ross H, Allen W, Ward J, Deffuant G, de Garine-Wichatitsky M, Binot A (2018) Systems thinking in practice: participatory modelling as a foundation for integrated approaches to health. Front Vet Sci 5:303.

Murphy KP (2012) Machine learning: a probabilistic perspective. In: Adaptive computation and machine learning series. MIT Press, USA

Neethirajan S (2017) Recent advances in wearable sensors for animal health management. Sens Biosensing Res 12:15–29

Robert CP (2014) Bayesian computational tools. Annu Rev Stat Appl 1:153–177. https://doi.org/10.1146/annurev-statistics-022513-115543Return to ref 31 in article

Suravajhala P, Kogelman LJA, Kadarmideen HN (2016) Multi-omic data integration and analysis using systems genomics approaches: methods and applications in animal production, health and welfare. Genet Sel Evol 48:38.

Index

D

S